For Joan and George

Table of Contents
Dr. Phil's Pocket Radio DX
©2016

This page intentionally blank.

1.1 Pocket Radio HF SDR: $18 DX
MW, SW, FM, AIR, VHF, UHF
©2015

WARNING: Wear eye protection when soldering.

1. HARDWARE: 24 MHz to 1.7 GHz

A NESDR Mini USB DVB-T & RTL-SDR receiver was purchased on eBay for $18.00 (shipped) from NooElec. A magnetic 4.75" antenna (diode protected; male MCX jack) was included. The NESDR Mini dongle can be used as an RTL-SDR. SDR refers to a software defined radio. RTL refers to the Realtek RTL2832U interface chip. The RTL2832U uses an ADC (Analog-to-Digital Converter) to accept an IF, low IF, or Zero-IF signal and convert it to 8-bit I/Q samples. The samples are sent via the universal serial bus (USB 2.0), at rates as high as 3.2 MS/s (mega-samples per second). The NESDR Mini's Rafael Micro R820T tuner chip covers from 24 to 1766 MHz. It consists of a low noise amplifier (LNA), I/Q mixers, and I/Q amplifiers. A 28.800 MHz crystal is utilized by both chips.

2. HARDWARE MODIFICATION: DC to 28.8 MHz

This mod takes soldering one piece of wire. The NESDR's plastic shell can be pried apart using a jewelers screwdriver. A wire was tied to a PCB hole (picture below) to prevent damage from pulling. One side of the wire was soldered to a circular, copper pad near the RTL2832U chip. The pad (Q+) is located between the markings C33 and L5 (picture above). The other side of the wire was fished through an existing case hole (see top picture). The wire was hooked to a 50 foot indoor antenna via an alligator clip. **A safer setup**: use a 10 nF capacitor to block DC and back-to-back diodes (to ground) to dump voltages over 0.7V. *Elsewhere* see other ways of presenting RF to the chip: ex. via a toroid. Ideally the other pad would be connected to ground using a 10 nF capacitor.

3. SOFTWARE

Stations from DC to 1766 MHz, including MW, SW, and FM DX, can be heard using software. Popular programs include: GQRX (Linux), SDR# (Windows), and SDR Touch (Android). An Android-OS based phone or tablet can use the NESDR Mini via a USB-to-microUSB cable and SDR Touch.

4. GQRX: UHF/VHF

GQRX utilizes rtl-sdr (free SDR) and gnuradio (free signal processing). From the terminal type: "sudo rmmod dvb_usb_rtl28xxu" each time the dongle is plugged into a USB slot. This uses root privilege to unload an auto-loaded module that will tie up the tuner and prevent its usage. From the terminal type: "gqrx" to start GQRX. To hear **UHF/VHF** set Device to Realtek RTL2832U; set Device string to rtl=0; set Sample rate to 2400000; set Bandwidth to 1 MHz; and set LNB LO to 0 MHz. The LNB LO is the low-noise block down-converter local oscillator. Use manual LNA gain. Proper manual LNA gain adjustment can be critical to VHF and UHF reception. The modulation used on VHF/UHF is often NFM (narrow-FM): WFM (wide-FM: mono or stereo) and AM are also used.

5. GQRX: MW/SW and MW PANADAPTER

After hardware modification, GQRX can be used as a MW and SW receiver using the direct sampling mode: input number **four**. For antennas, the outer metal ring of the MCX jack can be used as ground. To see the **entire MW spectrum** (from 448 to 1781 kHz); set Device to Other... ; set Device string to rtl=0,**direct_samp=4**; set Sample rate to 1500000; set Bandwidth to 1 MHz; and set LNB LO to 0 MHz. Enter an offset (upper right) of zero (0.000 kHz) to allow direct frequency entry. Enter a hardware frequency (upper left) of 1.115 MHz; the exact center of the MW band.

Turn the radio on by pressing the "Start DSP Processing" circle (upper left). Set the filter bandwidth: normal will suffice. Set mode to AM. Set AGC to fast for scanning (test medium for listening). Set the audio gain (volume). Adjusting the volume can be crucial for proper reception. Use the channel filter offset to change MW frequencies. Click the upper (or lower) part of the 10-kHz (XX#0.000 kHz) offset digit to increase (or decrease) the received MW frequency. Under "Input controls" set the LNA gain to zero and check "DC rem" (automatic DC removal). Before tuning on MW, center one station via the channel filter offset in kHz and tenths of kHz (XXX#.#00 kHz). The whole MW spectrum can be seen. The FFT size (display crispness) and rate can be set. Up to 3.2 Mhz of bandwidth can be seen at once. "No limits" may help switch into the direct sampling mode.

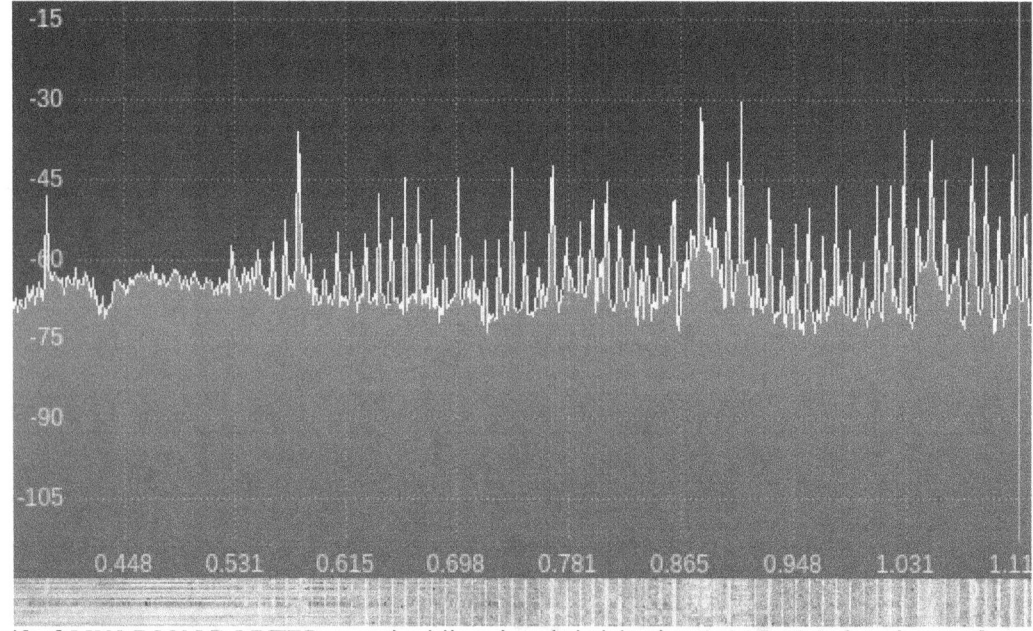

Left half of **MW PANADAPTER**: vertical line (at right) is the 1.115 MHz hardware frequency. Spikes represent stations' carriers. The sidebands, beside each carrier, contain voice data.

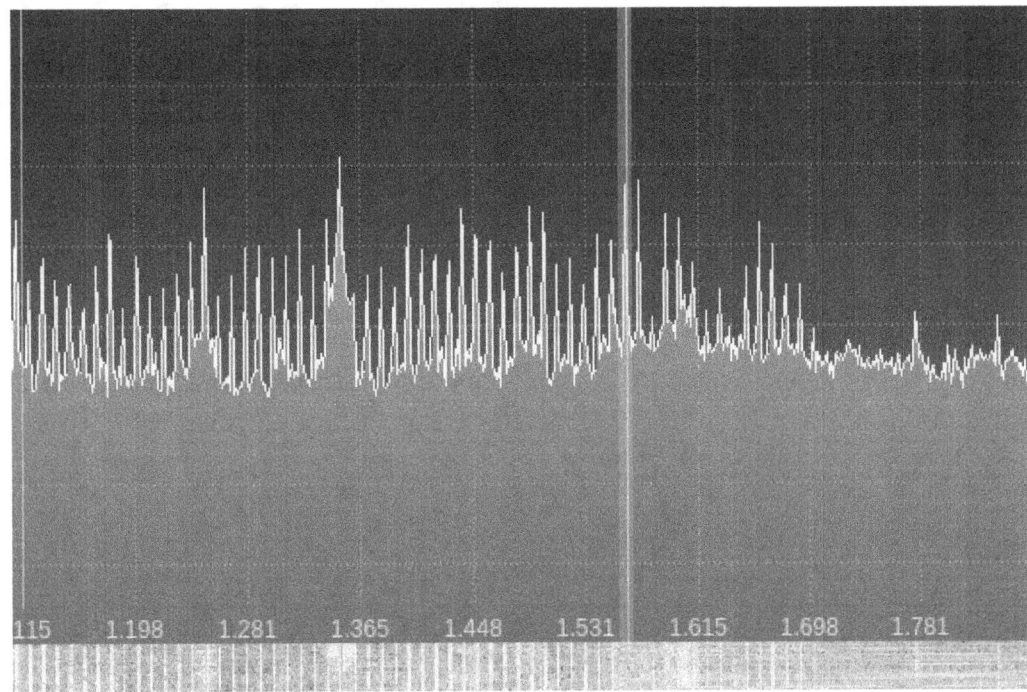

Right half of **MW PANADAPTER**: vertical line (above) is the station currently being heard. Clicking on the waterfall (bottom; larger on the actual screen) changes stations.

To hear **SHORTWAVE**, use the MW settings but set Sample rate to 2400000. Standard bandwidths on AM are 4.8-kHz narrow (*2.4-kHz voice*), 9.6-kHz normal (*4.8-kHz voice*), and 24.2-kHz wide (*12.1-kHz voice*). User defined bandwidths are: 0.5-kHz to 40-kHz on AM and 260-Hz to 9.8-kHz on SSB. GQRX will tune in fine 1-Hz increments; allowing quality reception of ham radio operators on SSB. Demodulation also includes CW-U and CW-L or continuous wave (Morse code). A "configuration" file can be saved for UHF/VHF-mode, MW-mode, and SW-mode. Restart the terminal and/or use "gqrx --reset" to switch between the MW/SW mode and the UHF/VHF mode.

6. POCKET RADIO

Pocket Radio DX was started in 2002 using an $11 Sony ICF-S10MK2 and RadioShack loop to MW DX. It evolved into using sub-$20 radios, almost disposable electronics, to MW, SW, and FM DX. Now an $18 **DC to 1.7 GHz Software Defined Radio** can be added to your Pocket Radio DX arsenal. And it can be used as a MW panadapter or to DX on VHF and UHF (scanner frequencies). The NooElec NESDR Mini tuner has the easy-solder, copper pads for direct sampling. It contains the RTL2832U and R820T chips. It is available for $18.00 on Ebay: item number 151217088428 ("Improved Capacitors & Crystal"). Or buy it directly from NooElec Inc. for $17.95 as SKU: 100556.

The modification in this article uses a clean input on the RTL2832U chip. This is exploited via using "**direct_samp=4**" in the device string. I recommend tinkering with both the software and hardware. Try tuned SW loops, tuned MW ferrite antennas, and inductor-based inputs using both the Q+ (*pin 4*) and Q- (*pin 5*) branches. Ideally, band-pass filters or traps would keep local MW and FM energies out of the LNA and ADC. Considering the 48-dB dynamic range of the RTL2832U's 8-bit ADC and its 28.8 MHz sampling rate, this is a surprisingly useful software defined radio. With SDR you can both hear and see radio frequencies. Since most of an AM station's energy is in its carrier, stations can be seen (above the noise floor) before they can be heard. MW, SW, FM, VHF, and UHF DXing with this "$18 version of the $2000 IC-R8500" is both challenging and fun.

Below are screen shots of the entire MW and SW spectrum: seen in seven viewings. Due to "folding", all the frequencies (up to 28.8 MHz) are seen by viewing the first 14.4 MHz of spectrum.

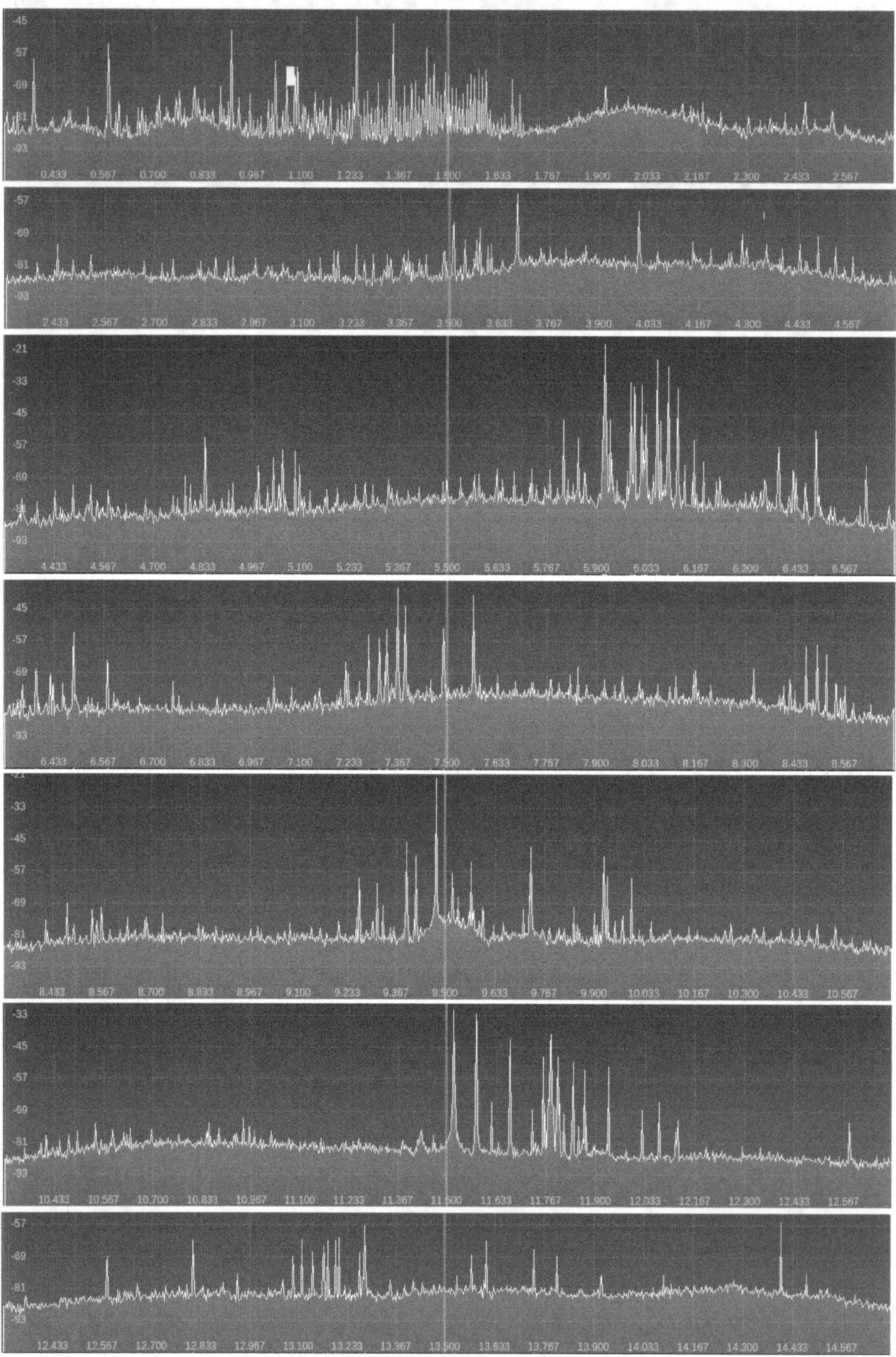

1.2 Pocket Radio HF SDR Cookbook
Direct Sampling Concepts
©2015

1. RATIOMETRIC CLOCKING

US patent 7,471,940 is titled "*Ratiometric clock systems for integrated receivers and associated methods*". This, Silicon Labs, patent describes dividing down a single oscillator for use by digital circuitry, including: ADC, DSP, MCU, and other clocked, digital components. The purpose is so that digital harmonics fall on the frequency of the oscillation signal (mixer LO). It may be advantageous to set the RTL2832U's sampling rate at frequencies divided down from its native 28.800 MHz, external clock. This may reduce noise over other sample rates (still being tested).

GQRX	Exact Sample Rate	Comment
250000	250000.000414 Hz	TOO LOW
1200000	-	RATIOMETRIC
1500000	1500000.014901 Hz	-
1700000	1700000.005629 Hz	-
2000000	2000000.052982 Hz	-
2200000	2200000.014570 Hz	-
2400000	-	RATIOMETRIC
2700000	2700000.160933 Hz	-
3200000	-	TOO HIGH

Above is a list of sample rates in GQRX: note that two are ratiometric. Sample rates (param samp_rate) can be set from 225,001 Hz to 300,000 Hz and 900,001 Hz to 3,200,000 Hz. The intermediate sample rates will cause choppy audio. Sample loss is expected above sample rates of 2.4 MS/s. Below is a list of ratiometric sample rates that are directly divided down from the 28.800 MHz crystal that clocks the RTL2832U. This keeps ADC sampling in line with the DSP and MCU clock cycles. Alteration of the sample rate can sometimes effect the intelligibility of a signal. A favorite sampling rate is 1,800,000 Hz. After more testing, the sample rates in GQRX, SDR#, SDR Touch, and other RTL-SDR programs may want to be changed to the ones in the left list below.

DIVISOR	RATIOMETRIC		DIVISOR	RATIOMETRIC
30	960000		125	230400
25	1152000		120	240000
24	1200000		100	288000
20	1440000		96	300000
18	1600000		-	-
16	1800000		-	-
15	1920000		-	-
12	2400000		-	-

2. IMAGES

Nighttime DX is concentrated from 5.7 MHz to 10.0 MHz. These frequencies are under 14.4 MHz, or half the frequency of the 28.800 MHz crystal oscillator. Daytime DX is concentrated from 11.5 to 17.9 MHz. At 0.0 MHz, 14.4 MHz, and 28.8 MHz artifacts can be seen and heard. Any station may be heard at an image located at the absolute value of the station of interest minus 28.800 MHz. For example: A 21.580 MHz station will show up again at |(21.580 - 28.800)| or 7.220 MHz. An image frequency typically sounds similar and is similar in strength. The MW band will show up, again, from 27.090 MHz to 28.280 MHz but in reverse order. Meaning, the station at 520 kHz will be heard at 28.280 MHz and the station at 1710 kHz will be heard at 27.090 MHz.

3. FM INTERFERENCE

Images from the *strongest and closest* broadcast FM band (88 to 108 MHz) stations will be present and can cause considerable interference. The images will be located at the absolute value of the frequency of the station minus whole number multiples of the 28.800 MHz clock speed. This includes both positive and negative values (whose absolute value is up to 28.8 MHz). Visit radio-locator, enter a zip code, press "go", and look for stations that are both physically close in miles and have a signal strength of 5 (as shown by the meter icons). Alternatively, reduce an FM radio's whip antenna and see what stations can still be received. Or step through 0 to 28.8 MHz using WFM and see what FM stations can be heard. These stations have the most energy. At my location the powerful stations are: 105.7 MHz (strong), 101.3 MHz (moderate), and 96.1 MHz (moderate). The strong station's interference spans ~200 kHz of spectrum; this is unacceptable. Below, the three interfering FM stations are shown in the chart with multiples of 28.8 MHz subtracted from the primary frequency (going from left to right). The columns are as follows: $f1 = -28.8 \times 1 = -28.8$ MHz; $f2 = -28.8 \times 2 = -57.6$ MHz; $f3 = -28.8 \times 3 = -86.4$ MHz; $f4 = -28.8 \times 4 = -115.4$ MHz. The strong station at 105.7 MHz will appear at any subtracted multiple of 28.8 MHz that is within the range of 28.8 to -28.8. Meaning, 105.7 MHz will interfere at 19.3 MHz and 9.5 MHz. And 101.3 MHz interferes at 14.9 MHz and 13.9 MHz. And 96.1 Mhz interferes at 9.7 MHz and 19.1 MHz.

MHz	f1	f2	f3	f4
105.7	76.9	48.1	19.3	-9.5
101.3	72.5	43.7	14.9	-13.9
96.1	67.3	38.5	9.7	-19.1

4. GAIN CONSIDERATIONS

Too much audio gain will cause distortion. This is gain within the demodulation program, not analog gain (the volume control) at a speaker amplifier. It is recommended that the "Hardware AGC" be left off. Activation of hardware AGC will increase both the noise floor and signal equally. This is not helpful. The "LNA gain" or low-noise amplifier gain will not affect the Pocket Radio SDR modification because the R820T chip is being bypassed. For MW/SW DX, leave the LNA gain at zero. For VHF/UHF DX, it is critical to adjust the LNA gain. With too little gain, the signal will be missing, in the noise floor. With too much gain, noise will swamp the signal. Attempts to add a BJT "common base" amplifier (MPSA18-based) or JFET "common gate" amplifier (J310-based) both failed to improve the signal-to-noise ratio. These topologies cause voltage amplification. Input impedance is low (an incoming space wave has an impedance of 377 ohms) and output impedance is high (the RTL2832U input is likely 15k-ohms). An ADC measures voltage: it converts a voltage into a number that corresponds to the amplitude of that voltage. Attempts to enhance voltage gains using toroidal turns ratios also failed. More experimentation is planned. The RTL2832U is sensitive "as-is". Reduction of noise may be more critical to enhancing reception.

5. ANTENNA

A fifty-foot, indoor, long-wire antenna was directly attached to the QP input (*pin 4*) of the RTL2832U. For protection from *direct current* add a capacitor (10 nF) and from *voltage* add back-to-back silicon diodes (to ground). An outdoor antenna would likely be a danger to the RTL2832U. The capacitor and diodes will not stop electrostatic discharge (ESD) from destroying the RTL2832U.

6. USB COMPUTER RISKS

To prevent fires, the universal serial bus specifications require over-current protection. The current limit cannot be above 5 amps. On startup, some devices draw high current due to capacitance. Each USB port must allow 1500 mA before current limiting. Even with this protection it is possible to burn your computer's motherboard via soldering on a USB device. Be careful and never solder while the device is plugged in. The PCB around the QP pin is typically a ground plane.

7. INTERFERENCE REDUCTION 1

The key to reception is the reduction of noise. The first step to reduce noise is an in-line RF choke (~100 uH): antenna to RF choke to radio. FM, UHF, and VHF can travel easily through even a small capacitance. But they are effectively blocked by inductance. A gimmick capacitor from the antenna feed to ground (or the foil, see below) can also be used to bleed off VHF and UHF energy.

8. INTERFERENCE REDUCTION 2

The second step to reducing noise is to enclose the tuner and all excess USB cable in aluminum foil (Faraday shield). Put excess USB cable in a loop, tie it, and then wrap in aluminum foil. A tiny box can be "gift wrapped" in aluminum foil using tape. The tuner and choke can be placed in the box. The key is ~360 degree foil enclosure. Avoid touching the foil to the input wire.

9. CLEANER SIGNAL

The noise reduction steps above resulted in the "noise floor" dropping from about -77 dB (gqrx) to about -84 dB. Even strong FM stations (25,000 Watts within miles) can be reduced to somewhat manageable interference using these measures. This will allow hearing signals originally masked by the immense energy of local FM stations. The radio will improve dramatically. The waterfall will change from white to deep blue (in GQRX). And CW signals can easily be seen in the waterfall. Below is the tuner in a small, foil wrapped, box with an antenna alligator-clipped to the RF choke. USB extension cable foil is not needed with a shielded cable (with two end toroids).

10. GQRX FILTERS

Like a diode, gnuradio will demodulate AM signals as long as the carrier is somewhere in the filter bandpass. If two carriers are present, the one with the largest amplitude will be used for demodulation and the other carrier appears suppressed. **Sideband-selected AM** can be done. Select the narrow filter and offset tune so that only the carrier and the cleanest sideband is in the passband. This, along with a medium or slow AGC, may reduce the distortion associated with selective fading. The filters under GQRX are razor sharp. If a second carrier is just outside the passband it will not be heard. With similar sized analog filters, a tone would surely be present. The "normal" 9.6-kHz filter is like a "perfect" 4.8-kHz (voice) analog filter, with a shape factor of one. Selective fading distortion can also be reduced using ECSS. **ECSS** is the tuning of AM (double-sideband, carrier, amplitude modulated) signals using SSB (single-sideband). Tune in 10-Hz steps until USB and LSB sound similar; try using the wide filter. This works better on an SDR than on other radios due to the sharp digital filters. Listen to the sideband with the least interference. The GQRX squelch can be used to keep noise to a minimum. After tuning to a dead spot on the frequency range of interest, press the "A" to automatically squelch to the current noise level.

11. GQRX AGC

In GQRX, the fast, medium, and slow AGC (Automatic Gain Control) all have a threshold of -110 dB (1.00 µVp, superb), a slope of 2 dB, and decays of 100, 800, and 2000 mS, respectively. Hang is disabled. The AGC of the RTL2832U chip is quite effective. However, it is possible to turn off the AGC and adjust the gain manually. Note that a gain of 0 dB to 100 dB is possible. Unlike analog radios, the fast AGC setting in GQRX works surprisingly well for program listening. By stretching out the "User" AGC input box, precise values (*threshold*, *slope*, and *decay)* can be tried.

12. RTL API

The RTL API (Application Programming Interface) allows an RTL2832U dongle to be a SDR. It can get a USB device string (rtlsdr_get_device_usb_strings); which includes: index, manufacturer name, product name, and serial number. Mine returns "*Using device #0 Realtek RTL2838UHIDIR SN: 00000001*". The API can open (rtlsdr_open) or close (rtlsdr_close) the RTL2832U device. And get the tuner type (rtlsdr_get_tuner_type: RTLSDR_TUNER_R820T = 5, RTLSDR_TUNER_E4000 = 1). It can set the sample rate (rtlsdr_set_sample_rate). And enable the direct sampling mode (rtlsdr_set_direct_sampling). Under direct sampling, the RTL2832U *IF-mode* is activated. The API can set the center frequency (rtlsdr_set_center_freq), that will control the IF-frequency of the DDC. This allows tuning from 0.0 MHz to 28.8 MHz. The API can enable offset tuning (rtlsdr_set_offset_tuning) to avoid DC offset and 1/f noise. API data streaming functions include: buffer reset (rtlsdr_reset_buffer), read samples synchronously (rtlsdr_read_sync), read samples asynchronously (rtlsdr_read_async), and cancel asynchronous operation (rtlsdr_cancel_async).

13. DIRECT SAMPLING MODE

The RTL2832U typically demodulates digital video signals and sends an MPEG stream via USB. Researchers found (by sniffing) that, under Windows, raw samples were sent to the PC for demodulation. Using an undocumented register, the RTL2832U can be asked to send unprocessed, 8-bit, baseband samples via USB. These are then processed by gnuradio. The specific register writes used to initiate the direct sampling mode are shown below. The program will respond with: "*Enabled direct sampling mode, **input 4***" . The C code is by Steve Markgraf and Dimitri Stolnikov.

```
rtlsdr_demod_write_reg(dev, 1, 0xb1, 0x1a, 1);   // disable zero-IF mode  DVBT_EN_BBIN
rtlsdr_demod_write_reg(dev, 1, 0x15, 0x00, 1);   // disable spectrum inversion  DVBT_SPEC_INV
rtlsdr_demod_write_reg(dev, 0, 0x08, 0x4d, 1);   // enable I ADC input  DVBT_AD_EN_REG
rtlsdr_demod_write_reg(dev, 0, 0x06, 0x90, 1);   // swap I and Q ADC  DVBT_OPT_ADC_IQ
dev->direct_sampling = on;                       // direct sampling set
```

The function rtlsdr_demod_write_reg uses the function libusb_control_transfer. The five passed values are: device, page, address, value, and length. The values of dev (0) and on (4) are passed via the function call. Which are passed from GQRX's "**Device string**" of "rtl=0,**direct_samp=4**". When rtlsdr_set_center_freq is called in the direct sampling mode, a call is made to rtlsdr_set_if_freq. This sets the frequency via three calls to rtlsdr_demod_write_reg (page 1, addresses 0x19, 0x1a, and 0x1b). If the function rtlsdr_set_offset_tuning is called in the direct sampling mode, an error is returned. Tuning is done with the DDC (Digital Down-Converter). A DDC produces a complex signal at zero frequency (baseband) from a digitized, real signal at an intermediate frequency. The SDR mode is enabled during rtlsdr_init_baseband via the call rtlsdr_demod_write_reg(dev, 0, 0x19, 0x05, 1); which (quoting source) "*disables DAGC (bit 5)*".

14. POCKET RADIO HF SDR

See Pocket Radio HF SDR for details on how to receive MW and SW using no up-converter, no fancy metal case, no balun, and no RF cables. Pocket Radio SDR, covering DC to 1.7 GHz, allows hearing DX on: MW, SW, FM, CB, HAM, CORDLESS, RC, NOAA WEATHER, TV, MILITARY, AIR, SHIP, RAIL, TRUNKING SYSTEMS, GPS, SATELLITE, and more. Pocket Radio DX was started in 2002 using an $11 Sony ICF-S10MK2 and a RadioShack loop to MW DX. It evolved into using sub-$20 radios, almost disposable electronics, to MW, SW, and FM DX. Now, via direct sampling, you can add a $18 **DC to 1.7 GHz Software Defined Radio** to your Pocket Radio DX arsenal.

1.3 RTL-SDR AM Images
Direct Sampling MW Image Frequencies
©2015

AM	Image	AM	Image	AM	Image	AM	Image
520	28,280	830	27,970	1140	27,660	1450	27,350
530	28,270	840	27,960	1150	27,650	1460	27,340
540	28,260	850	27,950	1160	27,640	1470	27,330
550	28,250	860	27,940	1170	27,630	1480	27,320
560	28,240	870	27,930	1180	27,620	1490	27,310
570	28,230	880	27,920	1190	27,610	1500	27,300
580	28,220	890	27,910	1200	27,600	1510	27,290
590	28,210	900	27,900	1210	27,590	1520	27,280
600	28,200	910	27,890	1220	27,580	1530	27,270
610	28,190	920	27,880	1230	27,570	1540	27,260
620	28,180	930	27,870	1240	27,560	1550	27,250
630	28,170	940	27,860	1250	27,550	1560	27,240
640	28,160	950	27,850	1260	27,540	1570	27,230
650	28,150	960	27,840	1270	27,530	1580	27,220
660	28,140	970	27,830	1280	27,520	1590	27,210
670	28,130	980	27,820	1290	27,510	1600	27,200
680	28,120	990	27,810	1300	27,500	1610	27,190
690	28,110	1000	27,800	1310	27,490	1620	27,180
700	28,100	1010	27,790	1320	27,480	1630	27,170
710	28,090	1020	27,780	1330	27,470	1640	27,160
720	28,080	1030	27,770	1340	27,460	1650	27,150
730	28,070	1040	27,760	1350	27,450	1660	27,140
740	28,060	1050	27,750	1360	27,440	1670	27,130
750	28,050	1060	27,740	1370	27,430	1680	27,120
760	28,040	1070	27,730	1380	27,420	1690	27,110
770	28,030	1080	27,720	1390	27,410	1700	27,100
780	28,020	1090	27,710	1400	27,400	1710	27,090
790	28,010	1100	27,700	1410	27,390		
800	28,000	1110	27,690	1420	27,380		
810	27,990	1120	27,680	1430	27,370		
820	27,980	1130	27,670	1440	27,360		

MW images are located from 27,090 kHz to 28,280 kHz.

1.4 RTL-SDR SW Images
Direct Sampling SW Image Frequencies
©2015

Band	kHz	kHz	Time	Image	Image	Image	0.0 to 14.4 Mhz	14.4 to 28.8 MHz
120M	2300	2495	night	26305	26500	clear	night	--
90M	3200	3400	night	25400	25600	clear	night	--
75M	3900	4050	night	24750	24900	clear	night	--
60M	4750	5100	night	23700	24050	clear	night	--
49M	5730	6300	night	22500	23070	clear	night	--
41M	6890	7600	night	21200	21910	13M	night	--
31M	9250	10005	night	18795	19550	15M	night	--
25M	11500	12200	day	16600	17300	clear	--	19M-16M
22M	13570	13870	day	14930	15230	clear	--	22M-19M
19M	15005	15825	day	12975	13795	clear	--	day
16M	17480	17900	day	10900	11320	clear	--	day
15M	18900	19020	day	9780	9900	31M	--	day
13M	21450	21850	day	6950	7350	41M	--	day
11M	25670	26100	day	2700	3130	clear	--	day

At night, all important shortwave band frequencies can be heard below 14.4 MHz. The only theoretical image interference is from 13M and 15M; which primarily cannot be heard at night. During the day, listening above 14.4 MHz, the 25M band appears between the 19M and 16M band. And the 22M band appears between the 22M and 19M band (in a similar position). The theoretical clash between the 41M/31M and 15M/13M bands does not happen because the former are nighttime bands and the latter daytime bands. The RTL2832U can direct sample up to 28.8 MHz.

1.5 RTL-SDR FM Images
Direct Sampling FM Image Frequencies
©2015

FM	f3	|f4|	FM	f3	|f4|	FM	f3	|f4|	FM	f3	|f4|
87.9	1.5	27.3	93.1	6.7	22.1	98.3	11.9	16.9	103.3	16.9	11.9
88.1	1.7	27.1	93.3	6.9	21.9	98.5	12.1	16.7	103.5	17.1	11.7
88.3	1.9	26.9	93.5	7.1	21.7	98.7	12.3	16.5	103.7	17.3	11.5
88.5	2.1	26.7	93.7	7.3	21.5	98.9	12.5	16.3	103.9	17.5	11.3
88.7	2.3	26.5	93.9	7.5	21.3	99.1	12.7	16.1	104.1	17.7	11.1
88.9	2.5	26.3	94.1	7.7	21.1	99.3	12.9	15.9	104.3	17.9	10.9
89.1	2.7	26.1	94.3	7.9	20.9	99.5	13.1	15.7	104.5	18.1	10.7
89.3	2.9	25.9	94.5	8.1	20.7	99.7	13.3	15.5	104.7	18.3	10.5
89.5	3.1	25.7	94.7	8.3	20.5	99.9	13.5	15.3	104.9	18.5	10.3
89.7	3.3	25.5	94.9	8.5	20.3	100.1	13.7	15.1	105.1	18.7	10.1
89.9	3.5	25.3	95.1	8.7	20.1	100.3	13.9	14.9	105.3	18.9	9.9
90.1	3.7	25.1	95.3	8.9	19.9	100.5	14.1	14.7	105.5	19.1	9.7
90.3	3.9	24.9	95.5	9.1	19.7	100.7	14.3	14.5	105.7	19.3	9.5
90.5	4.1	24.7	95.7	9.3	19.5	100.9	14.5	14.3	105.9	19.5	9.3
90.7	4.3	24.5	95.9	9.5	19.3	101.1	14.7	14.1	106.1	19.7	9.1
90.9	4.5	24.3	96.1	9.7	19.1	101.3	14.9	13.9	106.3	19.9	8.9
91.1	4.7	24.1	96.3	9.9	18.9	101.5	15.1	13.7	106.5	20.1	8.7
91.3	4.9	23.9	96.5	10.1	18.7	101.7	15.3	13.5	106.7	20.3	8.5
91.5	5.1	23.7	96.7	10.3	18.5	101.9	15.5	13.3	106.9	20.5	8.3
91.7	5.3	23.5	96.9	10.5	18.3	102.1	15.7	13.1	107.1	20.7	8.1
91.9	5.5	23.3	97.1	10.7	18.1	102.3	15.9	12.9	107.3	20.9	7.9
92.1	5.7	23.1	97.3	10.9	17.9	102.5	16.1	12.7	107.5	21.1	7.7
92.3	5.9	22.9	97.5	11.1	17.7	102.7	16.3	12.5	107.7	21.3	7.5
92.5	6.1	22.7	97.7	11.3	17.5	102.9	16.5	12.3	107.9	21.5	7.3
92.7	6.3	22.5	97.9	11.5	17.3	103.1	16.7	12.1			
92.9	6.5	22.3	98.1	11.7	17.1	103.3	16.9	11.9			

Interference will appear at the absolute value of the frequency minus whole number multiples of the 28.8 MHz clock rate that fall between zero and 28.8 MHz. Specifically of interest on FM: the FM frequency in MHz minus three times 28.8 MHz (**f3** column, -86.4 MHz) and the absolute value of the FM frequency in MHz minus four times 28.8 MHz (**\|f4\|** column, -115.2 MHz).

1.6 RTL-SDR Sample Rate Images
Bandwidth related images.
©2015

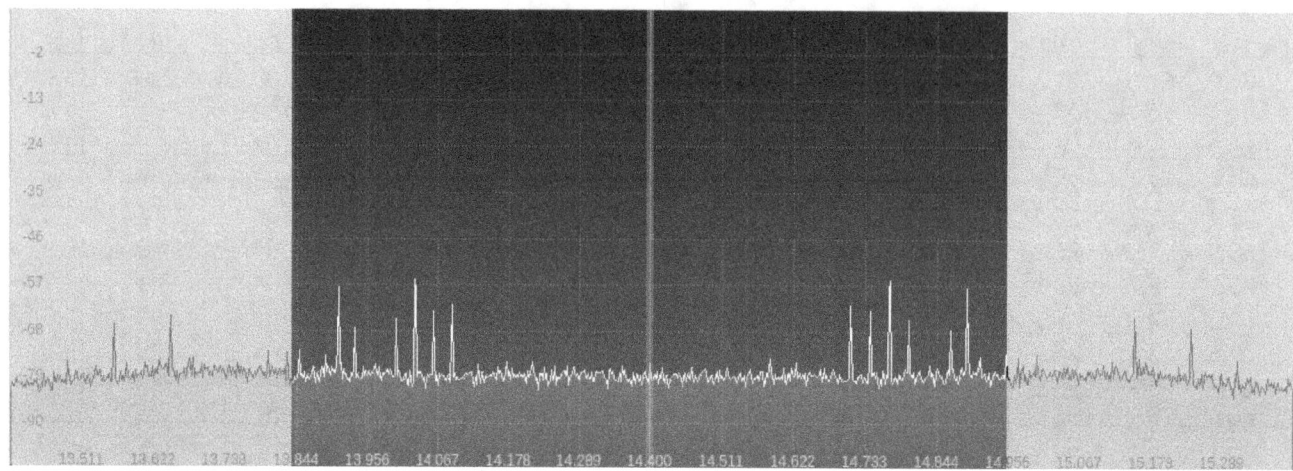

Another direct sampling image has been identified. It occurs with strong signals and is related to the sample rate. The "sample rate" is set to 2,000,000 Hz to view 2-MHz of spectrum. A strong signal at 9.840 MHz will be heard again at 9.840 MHz plus the sample rate or 11.840 MHz; but only at offsets to the far right of the screen (ex. +900 kHz). Due to folding, that 9.840 MHz signal can also be heard at the absolute value of 9.840 MHz minus 28.800 MHz or 18.960 MHz. This 18.960 MHz image signal will be seen again at 18.960 MHz minus the sample rate or 16.960 MHz; but only at offsets to the far left of the screen (ex. -900 kHz). These are sample-rate images.

A strong MW station, 1250 kHz, was tuned (-33 dB) at a 2.4-MHz sample rate. The offset was set to the far right: to +1080 kHz. The radio was tuned to the station plus the sample-rate: 3.650 Mhz (1.250 MHz plus 2.400 MHz). A -50 dB image was heard. At an offset of +980 kHz, the image at 3.650 MHz was -60 dB. And at an offset of +880 kHz, the image was in the noise floor.

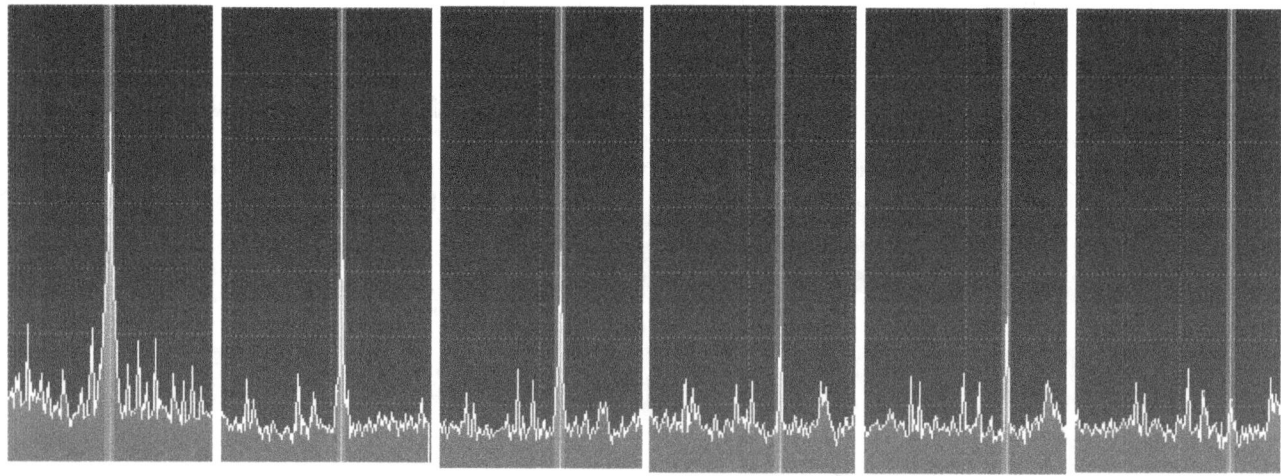

Above is an original signal and its 900, 800, 700, 600, and 500 kHz offset sample rate image. Any weak signal heard in the peripheral blocks, cannot be trusted as far as frequency. A station heard at the periphery that disappears when center tuned is likely a sample rate image.

Note: If the audio sounds distorted, make sure to turn down the DSP audio gain setting. Too much audio amplification will sound similar to what signal overload sounds like on an analog radio.

1.7 Realtek RTL2832U
The mystery chip at the heart of RTL-SDR.
©2015

1. INTRODUCTION

The Realtek RTL2832U is a 6 mm square chip (QFN) with 12 pins per side. It uses 3.3 Volts and a 28.8 MHz crystal. The chip's datasheet comes with a non-disclosure agreement. The IC is a DVB-T COFDM Demodulator. DVB-T stands for *Digital Video Broadcasting: Terrestrial* and COFDM stands for *Coded Orthogonal Frequency-Division Multiplexing*. The "U" in the name stands for the USB (2.0) interface. USB uses 4 pins: +Voltage, D-, D+, and ground. A USB extension cord is recommended with RTL-SDR to keep the radio away from computer generated noise. The RTL2832U contains eight general purpose input/output ports and an infrared remote control port.

2. TUNERS

The RTL2832U supports tuners at IF, low-IF, and zero-IF (direct conversion). The most common tuner is the Rafael Micro R820T. This chip has an advantage over the Elonics E4000 in that it only uses two of the RTL2832U's four I/Q pins. The unused pins, if not grounded, can be used as clean inputs for direct sampling purposes. The unused pins are the Q+ and Q- branches.

3. RTL_TEST

In Linux, using the terminal, "rtl_test" can be used to test the RTL2832U (Ctrl-Z exits). Program switches include: -s *sample rate* (the default is 2,048,000 Hz), -d *device index*, -t *Elonics E4000 tuner*, -p *ppm error measurement*, -b *output block size*, and -S *force synchronous output*.

4. DATA THROUGHPUT

A high-end sound card will push 32-bits at 192-kHz, a data throughput of 6.144 Mbps. The RTL2832U has only an 8-bit ADC but can run at 3.2 MS/s (million samples per second, 3200-kHz). This calculates out to a throughput of 25.600 Mbps or four times that of a good sound card. The RTL2832U was made to handle compressed MPEG2 and MPEG4 (H.264) transport streams; including video, audio, and data. Modulation schemes include: 4QAM (QPSK), 16QAM, or 64QAM. In 2k mode (1705 carriers), the carrier spacing is 4464 Hz. In 8k mode (6817 carriers), the carrier spacing is 2232 Hz. Under both modes the sampling rate is 18.284544 MHz and the final LO frequency is a fourth of that or 4.571136 Mhz (low-IF). Common bandwidths are 6, 7, or 8 MHz.

5. IMPORTANT PINOUTS

The important RTL2832U pinouts are as follows: *pins 1/2* In-phase Input pos/neg, *pins 4/5* Quadrature Input pos/neg, *pins 11/12* 28.800 MHz Crystal Oscillator, *pin 13* Tuner AGC, *pins 16/17* SCL/SDA Tuner Serial clock/data, *pins 18/19* SCL/SDA 2k (256 byte) EEPROM Serial clock/data S24CS0, *pins 25/26* LDO 3.3V AMS1117, *pin 38* Infrared Port, and *pins 40/41* USB Data pos/neg.

6. INPUT IMPEDANCE

The input impedance of the I and Q lines of the RTL2832U is "unknown". However, much is known about the Elonics E4000 chip. It is a "Multi-standard CMOS Terrestrial RF Tuner" in a 5 mm square package (QFN) that has 8 pins per side. It was designed to interface directly to a digital demodulator. The E4000's analog IF outputs; specifically *pin 20* (I+, IVOUTP), *pin 19* (I-, IVOUTN), *pin 18* (Q+, QVOUTP), and *pin 17* (Q-, QVOUTN) are connected, via a capacitor, directly to pins on the RTL2832U. And these I/Q baseband outputs can have a differential peak-to-peak output voltage of 1000 mV. Meaning, the RTL2832U can handle +4 dBm. The E4000 outputs also see an output load of 15k-ohms and 10 pF. Also telling, the outputs' single ended output impedance is 250-ohms (R_{OUT}). A space-wave has an impedance of 377-ohms. Therefore, there may be no significant impedance mismatch between the RTL2832U's I/Q pins and a random wire antenna.

7. ELONICS E4000

The E4000's has a 4.5 dB (or better) noise figure and a 50-ohm RF input. Total gain is ~99 dB, which includes: 30 dB LNA gain, 12 dB mixer gain, and 57 dB IF gain. There are six stages of IF gain, each with a maximum, digitally-programmable gain of: 6 dB, 9 dB, 9 dB, 2 dB, 15 dB, and 15 dB, respectively. The E4000 can handle 2000 mV peak-to-peak or +10 dBm. IIP3 is +5 dBm. It is important to realize that bypassing the E4000 or R820T also bypasses a huge amount of gain. Like radios of old, without RF amplification, use a long wire (SW), large ferrite (MW), or box loop (MW).

8. RTL2832U ARCHITECTURE: DIRECT SAMPLING

The heart of the RTL2832U is its ADC (Analog-to-Digital Converter) and DSP (Digital Signal Processor). It performs Digital Down-Conversion DDC (IF to baseband) via I/Q mixers (phase is 90 degrees apart), digital low-pass filtering, I/Q resamping, and sends 8-bit I/Q data via the USB port.

9. RTL2832U ARCHITECTURE: COFDM

Unused by RTL-SDR is the Fast Fourier Transform FFT unit. This converts time-domain information into frequency-domain information. Time-domain is when the y-axis is amplitude and the x-axis is time. Frequency-domain is when the y-axis is amplitude and the x-axis is frequency. I wrote a FFT algorithm for the PhilSCAN programs. The output seen on programs like GQRX, SDR#, and SDR Touch is frequency-domain information. Under direct sampling gnuradio does the FFT.

The RTL2832U is a COFDM demodulator, that does things like: symbol synchronization, fine frequency adjustment, phase rotation, channel estimation and correction, inner and outer deinterleaving, Viterbi decoding, RS decoding, forward error correction, adjacent and co-channel interference rejection, pre- and post- and long-echo channel reception, impulse noise cancellation, automatic carrier recovery, channel equalization, channel state information, guard period removal, pilot and TPS decoding, sample rate correction, sample rate interpolation and decimation, AGC delay, measurement of radio frequency levels, SNR estimation, control of the tuner's AGC, MPEG proportional integral derivative filtering, etc. The chip can automatically detect modulation parameters (ex. transmission mode, code rate, guard interval) via patent-pending algorithms.

10. POCKET RADIO HF SDR

See Pocket Radio HF SDR for details on how to receive MW and SW using no up-converter (~$43), no fancy metal case (~$24), no 4:1 balun (~$11), and no cables (~$10). DC to 1.7GHz coverage opens up a world of sub-$20 DX, including: MW, SW, FM, CB, HAM, CORDLESS, RC, NOAA WEATHER, TV, MILITARY, AIR, SHIP, RAIL, TRUNKING SYSTEMS, GPS, SATELLITE, and more.

1.8 Realtek RTL2832U Input
Secrets of the crab.
©2015

*The following is **speculative**.*

The public does not know the exact architecture of the Realtek RTL2832U. An 8-bit ADC was likely chosen to avoid the cost of a 12-bit, 14-bit, or 16-bit ADC and a similar bandwidth DSP or FPGA. The MCU inside the RTL2832U is 8051-based. During direct sampling, the tuner is bypassed, including its LNA, mixer, and IF gain. On an Elonics E4000, this is +99 dB of total gain. How is it possible for an 8-bit ADC in IF-mode to hear signals, directly from a long wire, without any gain?

During the call rtlsdr_init_baseband, the SDR mode is made possible by *"disable DAGC"*. This bypasses the COFDM demodulator. Yet another register write *"disable AGC"* is said to have *"no effect"*. During the call rtlsdr_set_direct_sampling, the Zero-IF mode is disabled. Realtek states that the IF-mode supports IF's of both 36.167 MHz and 36.125 MHz. The RTL2832U crystal runs at 28.8 Mhz, so reception is via folding. The 36.167 Mhz IF is seen at 7.367 MHz and 21.433 Mhz. The 36.125 MHz IF is seen at 7.325 MHz and 21.475 MHz. What else is being changed in the IF-mode?

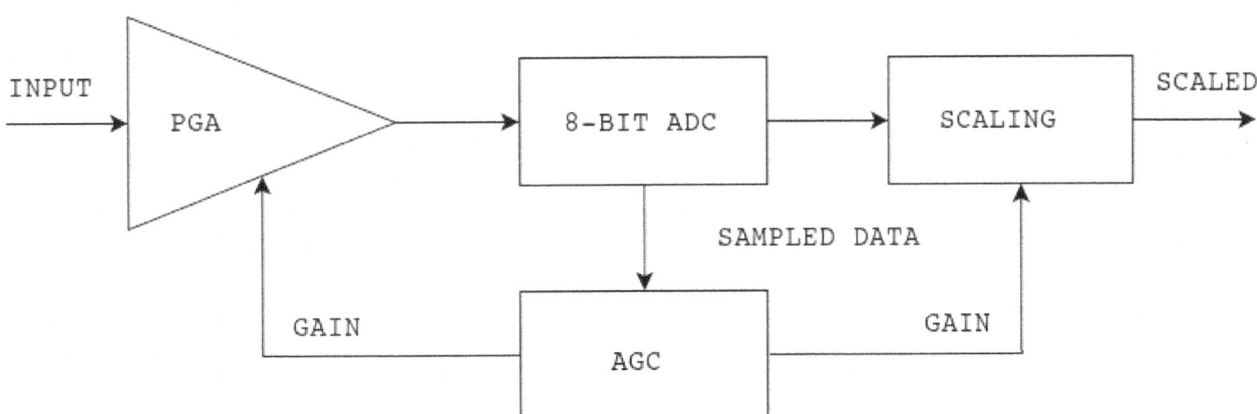

The digital AGC is presumed to only scale sampled values. I speculate that in the IF-mode an I or Q pin input is being sent to a programmable gain amplifier (PGA) controlled by an automatic gain control (AGC). The ADC digitizes the amplifier's output into sampled data. The AGC continuously adjusts the gain based on that sampled data and uses hysteresis to smoothly transition between gain levels. Glitches do not occur by switching gain during opportune times. Clipping is prevented by monitoring sampled peaks. The digital AGC can then scale the 8-bit number according to the applied gain. Discrete gain settings could be used: 1x, 2x, 4x, 8x, 16x, 32x. During direct sampling, it would be fruitless to scale the low-voltage, raw signal. A PGA-AGC system would increase the input-referred dynamic range (DR) of the 8-bit ADC. The chip may adapt to make the best use of its 48.1 dB DR. And results could trump a higher bit ADC solution. Realtek has produced AGC's that can set a PGA so that the input signal's peak-to-peak amplitude stays near the full-scale range of its 8-bit ADC. This ensures that the signal is digitized with the full resolution of the 8-bit ADC. Realtek's pre-ADC conditioning is advanced. Most SDR companies must use a high-bit ADC and achieve performance via "brute-force". They do not possess the complex calibration techniques and algorithms needed to pull this off using low-voltage semiconductors.

1.9 RTL2832U HF Direct Sampling
Setting Frequency and Sample Rate
©2015

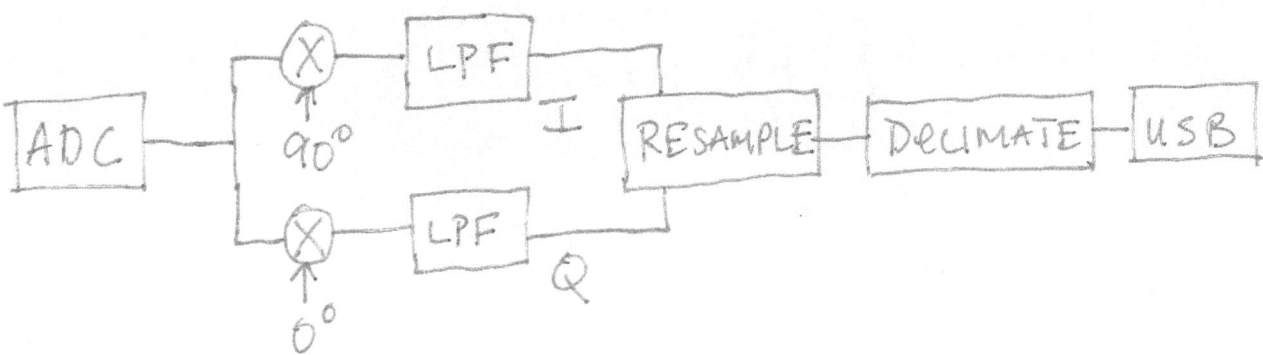

1. INTERMEDIATE FREQUENCY

After modification (see <u>Pocket Radio HF SDR</u>) an RTL2832U dongle can receive 0 to 28.8 MHz frequencies via direct sampling. The ADC (Analog-to-Digital Converter) sub-samples IF signals and a DDC (Digital Down Converter) converts them to zero frequency. The intermediate frequency is used to tune MW and SW frequencies, via the call rtlsdr_set_if_freq. This uses the register called DVBT_PSET_IFFREQ (22-bit: 21 MSB, 0 LSB) located on page 1, at addresses 0x19, 0x1a, and 0x1b.

```
IF_freq = IF_frequency_Hz % 28800000
```

The frequency is recalculated as the modulo or remainder of the division of the frequency by the clock frequency. This results in a value from 0 to 28,799,999 Hz. In the C programming language, the "%" operator is used for modulo math. It requires 3 bytes to hold a 22-bit number.

```
IF_freq = IF_freq * 0x400000 / 28800000 * -1
```

The frequency is multiplied by 2 raised to the power of 22, or 4,194,304, or 0x400000 in hex. And divided by the clock frequency (28.800 MHz) and multiplied by negative one. The last step is the two's compliment of the number. The MSB (most significant bit) is set when storing negative numbers: for 8-bits the number 1111 1111, which, unsigned, would represent 255, represents the two's compliment value of -1. And 1000 0000 is the two's compliment for -128.

```
IF_freq = IF_freq & 0x3fffff
```

By logically AND-ing with 0x3fffff hex, or 00111111 11111111 11111111 in binary, the first two bits are zeroed. This is a 22-bit number being stored in 3-bytes or 24-bits. Two bits will be zero.

```
rtlsdr_demod_write_reg(device, 1, 0x19, IF_freq >> 16, 1)
rtlsdr_demod_write_reg(device, 1, 0x1a, IF_freq >>  8, 1)
rtlsdr_demod_write_reg(device, 1, 0x1b, IF_freq       , 1)
```

The call rtlsdr_demod_write_reg is then used to write the three registers. The first call rotates IF_freq to the right (C operator ">>") by sixteen bits (2 bytes). This exposes the most significant byte. The second call rotates IF_freq to the right by 1 byte (8 bits) to expose the middle byte. And the third call (with no rotation) places IF_freq's least significant byte into the register. Device is a device handle for the RTL2832U chip. The first "1" in the code signifies page 1. The addresses 0x19, 0x1a, and 0x1b are consecutive in hex. The last "1" signifies that 1 byte is sent.

2. SAMPLE RATE

The other important parameter under direct sampling is the sample rate. The sample rate is set via the call rtlsdr_set_sample_rate. Valid values for the sample rate are from 225,001 Hz to 300,000 Hz and from 900,001 Hz to 3,200,000 Hz. It is possible to lose data at the higher rates. The intermediate rates will cause choppy audio. The sample rate uses a 26-bit register (27 MSB, 2 LSB) called DVBT_RSAMP_RATIO, located on page 1, at addresses 0x9f, 0xa0, 0xa1, and 0xa2.

```
sample_ratio = 28800000 * 0x400000 / sample_rate_Hz
sample_ratio = sample_ratio & 0xfffffffc
```

The clock frequency (28,800,000 Hz) is multiplied by 0x400000 hex and then divided by the sample rate, in Hertz. This becomes the sample ratio, that is then AND-ed with 0xfffffffc hex or 1111 1111 1111 1111 1111 1111 1111 1100. This has the effect of zeroing the two least significant bits: bits 0 and 1. The first significant bit, important to the register, is stored in bit 2.

```
rtlsdr_demod_write_reg(device, 1, 0x9f, samp_ratio >> 16, 2)
rtlsdr_demod_write_reg(device, 1, 0xa1, samp_ratio      , 2)
```

The call rtlsdr_demod_write_reg is then used, twice, to write four register values. The first call, located on page 1, writes 2 bytes, at addresses 0x9f and 0xa0. The value of the sample ratio is rotated right by 2 bytes (16 bits) to expose the two most significant bytes. The second call, located on page 1, writes 2 bytes, at addresses 0xa1 and 0xa2. The value of the sample ratio is written directly as the two least significant bytes. A sample rate change is followed by a soft reset.

3. SOFT RESET

After the sample rate is set, the demodulator is software reset. This is done via the DVBT_SOFT_RST register, located at page 1, address 0x01. This is accomplished with the third bit.

```
rtlsdr_demod_write_reg(device, 1, 0x01, 0x14, 1)
rtlsdr_demod_write_reg(device, 1, 0x01, 0x10, 1)
```

The call rtlsdr_demod_write_reg is used to write to the same register, twice. The first value, 0x14, is the value 0001 0100. Notice that bit-3 is set. The second value, 0x10, is the value 0001 0000. Notice that bit-3 is now zero. This sequence causes a reset. This is why the sample rate is typically set on a separate dialog menu. It cannot be set on-the-fly like the intermediate frequency.

Each C or Java line above will end with a semicolon (;). The values sent to rtlsdr_demod_write_reg must be cast (type converted) properly. For example, the last line of code above, in Java, may look something like the following. The function will return a result code.

```
return_code = rtlsdr_demod_write_reg( (byte) device, (char) 0x01,
                                      (char) 0x10,   (byte) 1 );
```

2.1 Pocket Radio DX
Fourteen Years of Slaying Radio Myths
©2002-2015

1. ABSTRACT

In 2003, a paper on using an inexpensive Pocket Radio to MW DX was made public: (http://www.radiointel.com/review-sonys10mk2.htm). It received over 30,000 page views. This article seeks to clarify that early work; and how it relates to other research, which spans 14 years.

2. INTRODUCTION

In 2002, I was returning to the MW DX hobby. Some gurus spoke of the need for a Drake R8B and a beverage farm to DX. How many turned away from our hobby, due to a lack of funds or space. Some DXers even pushed $4,000 receivers: the RX-340 and WJ-8711A. Stuck paying school loans, funds were limited. Most hobbyists were left to hear a small percent, the table scraps, of what the gurus caught with their expensive rigs. Like many hobbyists, I simply believed this "fact".

3. MASTERS THESIS

I went to college at 16, became a doctor, and defended an unrelated masters. In becoming a published researcher, part of our training was in analysis and statistics. This thinking spilled over into the radio hobby. In 2002, the disparity in *stations identified* between a high and low end receiver was *quantified*. The low end radio was Sony's S10MK2: $10.55 shipped and with taxes.

4. SONY ICF-S10MK2 POCKET AM/FM RADIO

In 2001, I wrote some R75 papers; including 3 describing modifications. The R75 Cookbook ended up translated into a few languages. And the Fidelity Increase Mod got carried by Kiwa Electronics. In 2006, Passport to World Band Radio stated: *"Of the three tested Kiwa modifications for the popular ICOM IC-R75, the audio Hi-Fi is the most useful."* I wrote RS232 programs to control the R75 while analyzing signals via a sound card's ADC [using my own x86 assembly fast Fourier transforms]. This allowed, back in 2002, band scanning, similar to today's ETM: see PhilSCAN.

In 2002, I wrote a paper titled Sony ICF-S10MK2 Pocket AM/FM Radio. And was grateful to *Ulis, Russ,* and *Jay,* at RadioIntel, for allowing such a controversial paper. The introduction posed one simple question: *"Can this inexpensive radio be used to pursue the hobby of AM DX?"* The next sections covered the features and specs of the radio. It then attempted to explain the MW DX hobby in *just three paragraphs*; touching on reception, station groupings, QSL cards, and putting in plugs for WRTH, NRC (National Radio Club), and IRCA (International Radio Club of America).

5. MATERIALS AND METHODS

Above was the 2002 setup: a Sony ICF-S10MK2 with RadioShack Loop (on the left) versus an ICOM IC-R75 and Quantum QX Loop (on the right). Basically $31 versus $780; the largest disparity available to me. The R75 has 1.77 µV sensitivity; the Quantum QX has +40 dB of RF gain.

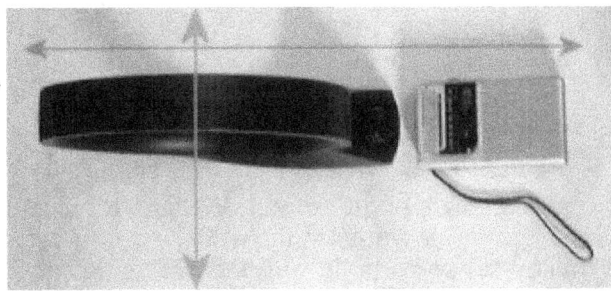

The Sony was tuned very slowly; as soon as one station faded, the loop was re-tuned, the distance between the loop and radio (coupling) optimized, and both were rotated through 180 degrees (loops being directional). The Sony sometimes needed additional fine-tuning. The picture above shows the positioning of the radio relative to the loop; as well as, the axis of reception (horizontal) and axis of attenuation ["nulling"] (vertical). The paper taped to the Sony was marked every 100-kHz, for use as a tuning aid. Earphones were used to make identifying stations easier.

6. RESULTS

Trial	Maximum	Neither ID	Both ID	R75 Only ID
1	113	30	70	13
2	113	26	80	7
3	113	14	98	1
4	113	25	85	3
5	113	8	94	11
6	113	30	80	3
7	113	17	90	6
8	113	0	112	1
Total	904	150	709	45

A maximum of 113 (not 119) stations, stems from the Sony tuning up to only 1650 kHz.

7. DISCUSSION

The Sony with RadioShack loop picked up **93.65%** of what the ICOM with Quantum QX loop did. Trial eight was a hot run, with two stations being heard on 19 frequencies. Meaning the Sony had identify 131 stations. The conclusion stated: *"The $11 Sony, **when coupled with a loop antenna**, is more than adequate for casual AM DX."* The word *"casual"* was added to prevent a flood of hate mail. Oddly enough, the article did not meet with harsh criticism. Nobody wants to hear that their expensive radio added less than 7% more identifications over an $11 Sony. More importantly, it was never the fault of the expensive radio; but merely **the fact that 93.65% of [my] MW [BCB] signals needed only a minimalistic AM radio in order to be identified**.

The Sony suffered from "*noise, bleed-over, images, overload, frequency drift*" and no frequency readout. However, it permitted an important DX benchmark: the ability to identify. The Sony does not suffer from MCU, synthesizer, LCD, or power line noise. And has a tuned frontend. Much MW DX depends on being in the right place at the right time: it is about just showing up.

The research proved the viability of using a low-cost, analog-tuned radio **with a loop** to MW DX. Suggested loops included: Torus, RadioShack, Terk, Select-A-Tenna, Edek, and home brew. Inexpensive radios have small ferrite rods and positively need larger BCB loops to compensate.

The S10MK2 contains a once revolutionary chip, the CXA1019S (see Advanced Low Voltage Single Chip Radio IC by *Taiwa Okanobu, Hitoshi Tomiyama*, and *Hiroshi Arimoto*). It also contains a quality tracking capacitor and ferrite rod, which form a tuned tank before its first RF amplifier. The Achilles heal is the small ferrite size; hence the use of the RadioShack loop. *It made no sense to test the stock radio*. As tested, Sony's tank forms half of a double-tuned inductively-coupled input. This is commonplace in crystal radio DX. Without the loop, the Sony would have done poorly. The CXA1019S chip is hardly an impediment compared to a crystal radio, which may rely on a double tuned input, followed by, say, a 5082-2835 Schottky diode, capacitor, and sound powered phones.

Which is more challenging: catching DX with a $4000 radio? An $11 radio? A single active device? A crystal radio? What makes radio fun is that DX can always be heard by using less.

Radio waves travel from miles away, yet some hobbyists feel the last few inches needed to be by wire. Variable inductive coupling has advantages. *It made no sense to replace the inductor of the stock radio's tank*. It makes sense to build one ferrite, or better, one 34 inch box loop and inductively couple it to the stock ferrite inside *any pocket radio*. This reduces antenna cost to that of an air-variable tuning capacitor. The MW band suffers from high noise [~10 dBµV, 3.16 µV, S3].

Extensive data analysis revealed important discoveries. About half (55%) of failures to identify were due to stations located *adjacent to* or *at the image of* the strongest local MW stations. Meaning that: living in an area with no local blowtorches, the expectation is to hear about 25 of those 45 "R75 Only ID" stations; or a whopping **97.28%** of what the modified R75 heard.

The other half (45%) of failures to identify were very weak signals. They were nearly impossible to hear on the R75: identification took time. On the Sony, these were simply missing "in the mud". Realize, in this 45%, there was never a weak signal that was easily heard on the R75 but missing on the Sony. What was missing on the Sony were some very rough stations to identify.

The high gain and low-IF image-rejection schemes of Sony's SRF-59 (analog, proprietary); as well as Kaito's KA321 and Radioshack's 12-586 (digital, Silicon Labs based) allow for potent Pocket Radio MW DX. Expensive radios are better; but, what *nobody had bothered to research* is that there is, often, only a small percent of signals where it matters, in identifying stations.

8. POCKET RADIO SW DX

To allow Pocket Radio SW DX, four papers of Sony Walkman modifications were introduced, including: S10MK2 to Shortwave, S10MK2 Handbook, SRF59 to Shortwave, and SRF59 Handbook. The SRF-59 retained its tuned input on shortwave! Techniques for soldering were written about in Soldering Lessons. Many hobbyists are hesitant to solder on a radio, even an $11 radio with large chip pins. Pocket Radio SW DX is now possible "off the shelf" with Kaito's KA321. An extensive (10-page) article, Hacking the KA321, includes various DX related modifications. In 2014, a homebrew radio, based on a $7 CXA1600P chip was introduced for SW DX; see Hellenized CXA1600P.

9. RADIO EDUCATION

I was taught RF electronics by another Greek, *Pete Gianakopoulos*, who has worked for both Rockwell-Collins and Motorola. Some favorite books include: Commmunications Receivers Princicples and Design by Ulrich L. Rhode and T.T.N. Bucher; Radio Engineers' Handbook by Terman; The ARRL Handbook 2004; and The Secrets of RF Circuit Design by Joseph J. Carr.

10. SYNCHRONOUS AM DETECTION

Passport to World Band Radio and its followers pushed the need for a Drake R8B to DX due to a feature called a synchronous AM (SAM) detector. I owned Drake-designed SAM detectors in the Satellit 800 and Eton E1. *Dallas Lankford* questioned the usefulness of SAM. In 2006, a paper called Tuning Tricks Challenge SAM described two methods from the 2002 R75 Cookbook, *applied to any radio*, that make SAM superfluous. The easiest, called sideband-selected AM was simple: select AM-mode, slow-AGC, and a narrow filter. Then offset tune by *nearly* half the filter's bandwidth (up or down, away from any interference). A boost in audio fidelity is noted. Old-timers likely, intuitively, used this method. The SAM feature herded many into buying costly receivers.

11. UPPING THE ANTE: AM

In 2007 the Angelodyne was released: the world's first regenerative plate detector. Helpful to the discovery was *Dave Schmarder* of TheRadioBoard and Dave's Homemade Radios. The first radio based on the detector, the Hellenedyne, uses one frame-grid, vacuum-tube triode (6GK5), 27 volts of plate (batteries), and 16-ohm earphones. Dave stated: *"This is a regenerative receiver using plate detection. This has become one of my favorite DX sets as the regeneration control is very smooth."* Professor *Arnaldo Coro*, an academic and host of Radio Havana, wrote: *"Thank you for sharing your know-how with the worldwide amateur radio hobby enthusiasts !!!"*

Over the years, buying guides were released: 2011 Shortwave Guide, 2008 Shortwave Radio Picks, 2006 Radio Buying Guide, 2005 Portable Guide, and 2005 Tabletop Guide; and a SSB Cookbook. And data on some favorite radios: KA2100 Superior by Design, DE1102 Mini-Cookbook, DE1103 Cheat Sheet, PL390 Cheat Sheet, Eton E1 SWL Tips, and FRG100 SWL Tips.

12. SILICON LABS

In 2010 several Silicon Labs related articles were released: Si4734 Hardware Guide, *Si4734 Software Guide*, Si4734 Homebrew Radio, Si4734 Offset Tuning, and Si4734 Digital Filters. Silicon Labs' low-IF, DSP receiver chips have revolutionized MW, SW, and FM. Their patents are amazing. Surprisingly, Silicon labs' lead engineer said that my articles were influential to the chip's design.

13. UPPING THE ANTE: FM

In 2011 the superiority of Silicon Labs' chips on FM was shown: PL390 FM DX King. Missing from the one active device (1AD) community was the ability to 1AD FM DX. In 2011, the Angelodyne Detector was morphed into the first single-transistor (JFET) FM radio built specifically for DX. The FM Heliosdyne logged 42 FM stations on its first night. It uses varactors and a 10-turn pot for effortless FM tuning. It uses 16-ohm phones instead of expensive, sound powered phones.

14. POCKET RADIO DX

The goal of Pocket Radio DX was to DX with radios under ~$20, nearly disposable electronics. The key is using a good antenna. This, basically, takes a crystal radio setup and replaces the diode with a cheap radio. In 2002, I realized that radio waves do not care about the cost or complexity of the radio used to hear them. Low-dollar, analog-tuned radios would suffice.

Recommended for MW Pocket Radio DX are the Sony ICF-S10MK2, Sony SRF-59, and *RadioShack 12-586* [Silicon Labs inside]. The frequency tuned on the 12-586 can be determined: see Hacking the KA321. Today's equivalent to the $20 RadioShack loop is a $40 Terk Advantage. A better match is a 34 inch box loop (see the NRC's research): inductively coupled to *any* MW radio.

Recommended for SW Pocket Radio DX are the modified Sony ICF-S10MK2, modified Sony SRF-59, and KA321 [Silicon Labs inside]. The KA321 was an exciting new radio, especially since most are hesitant to modify their radios. A tuned loop on shortwave, is useful for nulling out local noise. However, for DX, 25 to 50 feet of wire wrap near the ceiling suffices (capacitively coupled).

Radio	Maker	Cost	Target
ICF-S10MK2	Sony	$12.00	MW + SW (w/mods)
SRF-59	Sony	$17.00	MW + SW (w/mods)
12-586	RadioShack	$15.00	MW + FM
KA321	Kaito	$20.00	MW + FM + SW
NESDR mini	NooElec	$18.00	MW + FM + SW

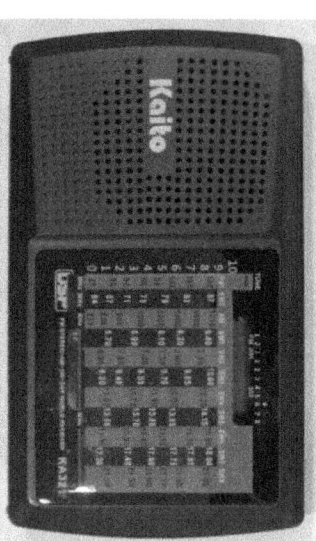

When were the MW DX capabilities of a Pocket Radio discovered? The original research on using a Pocket Radio to MW DX was done in 2002 and uploaded to *RadioIntel* in 2003. It is archived on February 17, 2004: see the *Wayback Machine*. The discovery was *not* made by accident.

15. SOFTWARE DEFINED RADIO

A new article (Pocket Radio HF SDR) adds a software defined radio to the Pocket Radio DX arsenal. A minor mod turns an $18 NooElec NESDR Mini DVB-T tuner into a MW/SW/FM DX radio.

16. CONCLUSION

In 2003, the expectation was for hobbyists to jump on the Pocket Radio DX bandwagon. In time, the tide finally turned. In 2011, the hope was to see a new hobby emerge: 1AD FM DX. That still has not materialized. Many hobbyists would like to see renewed interest in 1AD and crystal DX. Unfortunately, the leap to formulas and a soldering iron is a large one. Most MW or SW DXers would be shocked at what a single triode receiver, like the Hellenedyne, can hear. *Dave Schmarder* used his version of this radio in both the 2009 Radio Contest and the 2009 Active Device Contest.

I caught the radio bug at around 12, after my parents bought me a Six Million Dollar Man crystal radio backpack. My father, who collected radios in his youth, had a huge collection of Popular Science and Popular Mechanics magazines. I spent many hours, in our basement, sitting in front of a large bookshelf: dreaming about building the ultimate radio to DX with on MW and SW.

2.2 Sony ICF-S10MK2
FM/AM 2-Band Pocket Radio
⬜2003

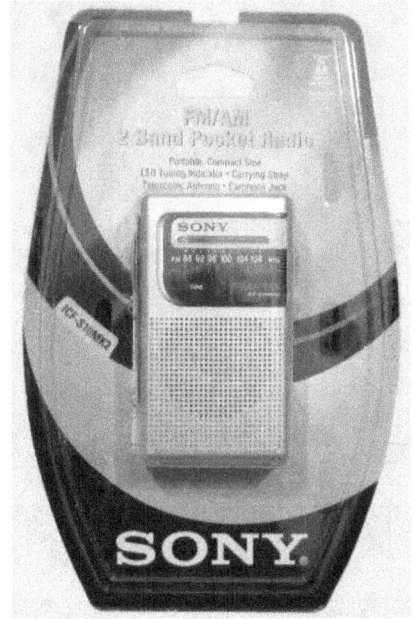 

This is the original 2003 article.

1. INTRODUCTION

The Sony ICF-S10MK2 is an "FM/AM 2-Band Pocket Radio" made in China. Cost including shipping and taxes is $10.55 when ordered directly from Sony's website. Can this inexpensive radio be used to pursue the hobby of AM distance reception (AM DX)?

2. FEATURES

The front includes a tuning scale, LED-tuning indicator, and built-in speaker. The right panel contains a tuning dial and AM/FM band switch. The left panel includes a power-switch volume-control wheel, earphone jack, carrying strap, and telescoping FM antenna. There is a rear battery compartment.

Note: The red tuning LED grows faint when the batteries need replacing.
Note: An internal ferrite bar is used for AM reception, not the telescoping antenna.

3. SPECIFICATIONS

1. Reception: 530 kHz to 1650 kHz AM and 87.5 MHz to 108.0 MHz FM.
A. Output: 100 mW at 10% THD via a 2.25" (5.7 cm) speaker.
B. Power: 3 Volts via 2 "AA" (R6) batteries, not included.
C. Battery Life: FM ~40 hours and AM ~45 hours.
D. Dimensions: 2.875" x 4.750" x 1.188" (71 mm x 119 mm x 30 mm) WHD.
E. Weight: 7 ounces (202 grams) with batteries.
F. Warranty: 1 year limited.

Note: Although Sony's literature states an AM tuning range of 530 kHz to 1710 kHz, testing revealed a range of 530 kHz to 1650 kHz.

4. THE AM DX HOBBY

The hobby of AM distance reception goes by two names: MW (mediumwave) DX and BCB (broadcast band) DX. The object is to identify (ID) and log stations using: time, call sign (usually given each half-hour), frequency, content, language, or local color (ex. weather or commercials).

Reception is best during the winter, decent in the fall, and poor in the summer. Daytime reception is poorer than nighttime reception. Nighttime offers less noise and some stations reduce or discontinue power. At local sunset stations to the west can be heard as stations in the east power down for the night, this is called "gray path" reception. Different catches will be made at sunrise, daytime, sunset, and nighttime. Station tests are often performed late on Sunday night. Conditions change minute by minute as stations fade in and out.

Clear channel stations with up to 50 kW power are located at: 540, 640-780, 800-900, 940, 990-1140, 1160-1220, and 1500-1580 kHz. Local channels (called the "graveyard") with 1 kW or less power are located at: 1230, 1240, 1340, 1400, 1450, and 1490 kHz. The graveyard contains many interfering signals. All other channels below 1610 kHz are considered regional, having 10 kW or less power. The "expanded AM band" runs from 1610 to 1710 kHz.

Some collect QSL (reception) cards. Simply mail the station a reception report consisting of the: date, time and time zone, frequency, program details (station ID, program name, host, commercials, etc.), and how well the signal was received (excellent, good, fair, poor). Make sure to include your name, address, and return postage! Making a recording of the broadcast will definitely help.

For station identification see WRTH (World Radio and TV Handbook), a yearly publication. Or visit the following websites:

> http://www.fcc.gov/mb/audio/amq.html
> http://www.radio-locator.com/

For more AM DX information including when stations perform DX and equipment tests join either the National Radio Club (NRC) or the International Radio Club of America (IRCA).

> http://www.nrcdxas.org/
> http://www.ircaonline.org/

5. PERFORMANCE

The Sony ICF-S10MK2 with a Radio Shack Loop was compared against a modified ICOM R75 with a Quantum QX Loop. This testing was, in essence, a shootout between a $31 portable setup and a $780 tabletop setup! The R75 provides ~1.77 µV AM sensitivity [PREAMP "1" ON] and the Quantum adds another +40 dB of RF gain.

Theoretically the Sony is capable of tuning to a maximum of 113 stations. Any station sounding good enough to allow identification was counted as being "heard". The chart below summarizes the data from eight separate comparison trials.

TRIAL	TOTAL	NEITHER HEARD	SONY and R75 HEARD
1	113	30	70
2	113	26	80
3	113	14	98
4	113	25	85
5	113	8	94
6	113	30	80
7	113	17	90
8	113	0	112

The results? On average the Sony/RS combo heard 94% of what the twenty-five times more costly ICOM/Quantum combo did. Impressive! Tuning through the AM band yielded an average catch of about 87 stations for the Sony. Trial eight was done on a particularly "hot" night and resulted in 112 stations heard on the Sony. What the chart does not show is that two distinct stations were resolved on 19 frequencies that night. This means that the Sony heard a total of 131 distinct stations including: news, business, talk, sports, religious, music, nostalgic, and ethnic. Incredible! The $11 Sony, when coupled with a loop antenna, is more than adequate for "casual" AM DX.

Note: The Sony was not penalized for the six stations it could not tune from 1160 kHz to 1710 kHz. In North America frequency spacing is 10 kHz, elsewhere 9 kHz. The ICOM maintained a higher signal-to-noise ratio than the Sony.

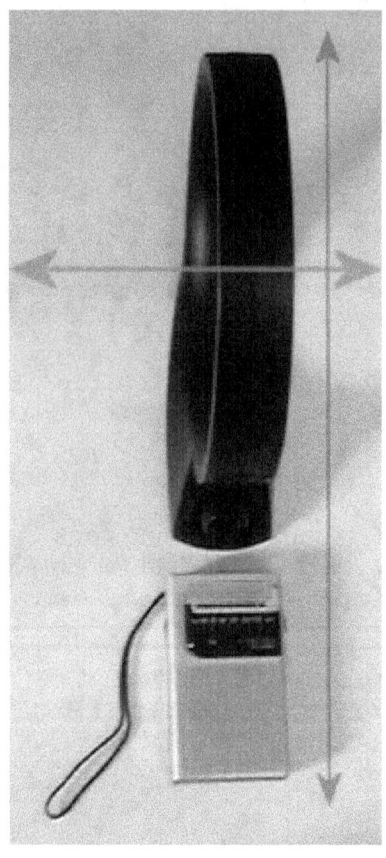

6. OPERATION

Using the Sony definitely takes some practice. Tune very slowly until the previous station just fades, then re-tune the loop antenna, and finally rotate the Sony/loop combo as a couple through 180°, as loops are highly directional in nature. The Sony may need additional fine-tuning. The picture above shows the proper alignment of the radio and antenna for reception along the vertical axis. Maximum attenuation ("nulling") of strong interfering stations occurs along horizontal axis. A red piece of paper was taped to the radio and marked every 100 kHz for usage as a tuning aid.

Note: Usage of headphones is recommended; always start with the volume low.

7. PROS AND CONS

The $11 Sony has several attributes going for it including: an analog local oscillator, a tuned loop input, and battery power (no AC hum problems). It is possible that the Sony ICF-S10MK2 uses a similar receiver chip and ferrite rod as their $90 digital portables. Ironically the more expensive and "flashy" digital Walkmans often experience processor and synthesizer related noise.

The Sony does suffer from noise, bleed-over, images, overload, and frequency drift. It is also difficult to know what frequency you're on using its analog tuning scale. Care must be taken not to miss stations, especially at the upper end of the tuning range. Data analysis revealed that stations adjacent to the strongest stations and stations located at the images of the strongest stations accounted for 55% of all failures to be heard. Much of the remainder was due to very weak signals.

8. CONCLUSION

Like to AM DX while on vacation? No money in the current radio budget for a "big gun" receiver? Need a truly portable performer? Want a radio suitable for usage where it may become lost, damaged, or stolen? Trying to introduce a child to AM DX? Then you may want to consider an $11 Sony ICF-S10MK2 and an inexpensive loop: Radio Shack, Torus, Terk, Select-A-Tenna, Edek, or homebrew.

9. HISTORY

In 1948 scientists at Bell Labs invented the transistor. Akio Morita purchased a license to build transistors for $25,000 in 1952. He quickly became the laughing stock of the business community. Bell Labs engineers informed him that transistors were "only good for making hearing aids". Akio Morita's vision was to manufacture transistor radios. Advisors warned him that: "radios are far too expensive to devote to one person".

In 1957 a company, co-founded by Morita, named *Tokyo Tsushin Kogyo Limited*, exported their six-transistor "*Sony*" TR-63. This $40 pocket-sized radio was so successful that the company changed their name to Sony and their motto to: "One Person, One Radio". This marked the start of Sony dominating the consumer electronics industry. It appears that after 50 years Sony remains true to their roots, producing high-quality pocket radios. Happy listening!

2.3 Hellenized ICF-S10MK2
$14 old-school radio hacker's portable
©2009

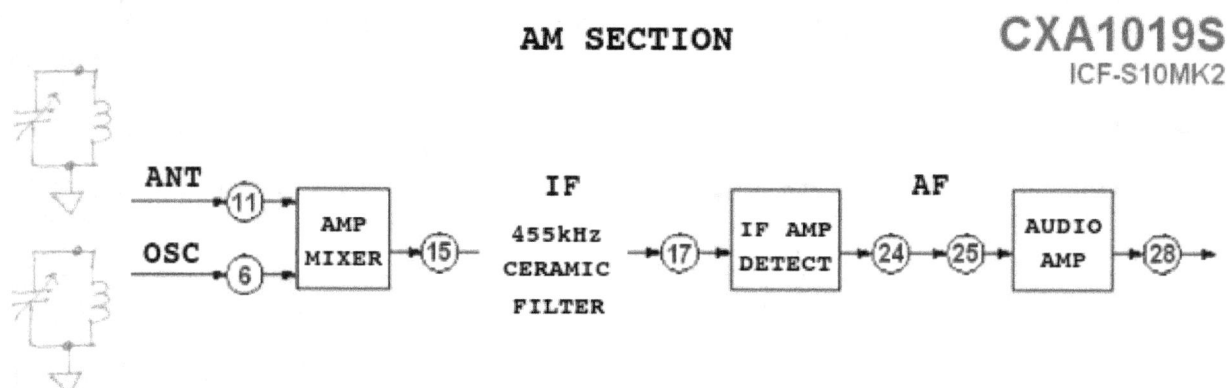

AM SECTION — CXA1019S (ICF-S10MK2)

1. INTRODUCTION

In 2003 I tested an $11 Sony S10MK2 with $20 Radio Shack Loop against a $580 ICOM R75 with $200 Quantum QX Loop. The former heard 94% of what the later could on BCB. So I became interested in the S10MK2's design and converted it to shortwave. The radio, based off of Sony's CXA1019S chip, proved to be an easy foundation for both AM and FM homebrew design.

2. CXA1019S

The Sony CXA1019S is a 2 Volt bipolar AM/FM radio IC. The chip includes an RF amplifier, RF AGC, oscillator, mixer, IF amplifier, IF AGC, detector, and AF power amplifier. It also contains volume control, tuning meter, and band-selection circuitry. AM mode current use is a low 3.5 mA. Both detector (0.6%) and audio (0.3%) distortion are low. Gain is distributed as follows: frontend 22 dB, AM IF ~53 dB, and audio 32 dB. The chip was revolutionary, incorporating everything that old-world "ceramic IF filter" radios needed to work. New chips are using I/Q mixer image rejection and either active RC filters (ex. CXA1129N in SRF-59) or DSP IF filtration (ex. Si4734 in DE1123). The CXA1019S's pins are small but large enough to solder on. This chip can be soldered on either side in an S10MK2. My method of soldering to tiny chip pins is described in **Phil's Soldering 101**.

CXA1019M
28-pin SOP

CXA1019S
30-pin SDIP
(S10MK2)

The CXA1019S is based on a 1.8 Volt chip design by *Taiwa Okanobu*. His 1982 prototype contained 700 elements; by contrast, the SRF-59's CXA1129N uses 1360 elements. The frontend design supports reception to 30 MHz and has a 2 dB noise figure. The mixer is a double-balanced type [level is +8.57 dBm] with 20 dB of conversion gain. The on-chip voltage regulator helps with power source and thermal-related fluctuations. The oscillator is isolated using N^+ diffusion islands. The AM IF (455 kHz) amplifier has band pass characteristics of: -10 dB at 1.25 MHz and -20 dB at 2.50 MHz. The detector has higher output than a standard diode, yet requires no adjustment. The audio amplifier includes pre-amplification, is of a complimentary push-pull variety, and has 0.18% THD. Volume control consists of up to -92 dB of attenuation. Sensitivity for 6 dB S/N is 0.63 µV or between S2 and S3; S/N is 50 dB. Although commonplace now, *Taiwa*'s design permitted for: 1) a reduction in external components to one-fifth, 2) low power requirements, 3) high sensitivity, and 4) audio output of ~500 mW. The IC's processing allowed for high hFE transistors, Si_3N_4 *thin film* capacitors, and two level wiring. *The pin-outs for the prototype IC match those of the CXA1019M.*

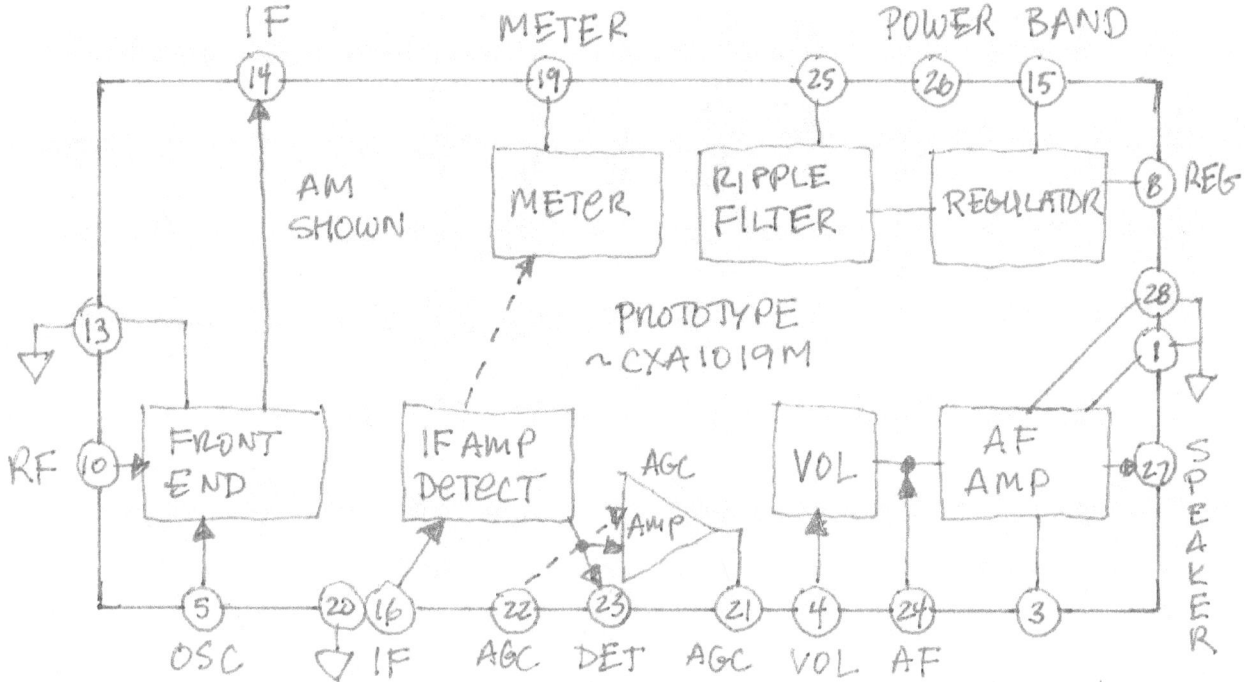

3. ICF-S10MK2

The ~$14 S10MK2 is great for radio hacking. The radio has a quality printed circuit board, durable case, good speaker, headphone jack, and an aerial antenna. The unit is small at 16.2 ci³ (cubic inches) yet other components can be shoehorned into the case. The analog tuning system is a plus: there is no CPU or synthesizer noise to contend with. Entire bands can quickly be tuned through and the powerful "listening" stations identified. Ironically this is a hassle on modern rigs. Two Duracell "D" type alkaline batteries *could* power the S10MK2 up to ~8 hours daily for a year.

4. HELLENIZED ICF-S10MK2

The S10MK2 must first be converted to receive shortwave stations. The line to *pin 6*, the local oscillator (LO), was severed using a pocket knife. The new LO tank inductor became an iron powder toroid (red; μi = 10). The LO tank capacitor became my trusty Heathkit Signal Generator, complete with 3:1 vernier drive and half-moon dial. The LO alteration needed two taps: *oscillator pin 6 and regulated power pin 9*. The **Q** is determined by the tank's inductor: Amidon toroids are good for a Q of about 250. As for the tuning range, I determined that about 90% of English night broadcasts reside between ~4.7 and 10.0 MHz [by hours of programming beamed to N. America].

The ferrite's tap was used to introduce shortwave energy. In this way the MW rod is intact and regulated voltage is applied to *pin 11*. The S10MK2 has adequate overall gain so no common-base amplifier was needed. A tuned loop proved to be valuable at attenuating local noise and had the added benefit of reducing various mixer artifacts. A longwire antenna resulted in more noise.

The S10MK2's anemic ceramic filter (tiny red rectangle near speaker cutout) was replaced with a Murata 15-element CFS455J with specs of -6 dB at 3 kHz and -80 dB at 9 kHz. Electrolytic capacitor size was increased on the AGC line. Using an offset tuned narrow filter with a slow-AGC reduces carrier dropout related distortion. This trick, **SIDEBAND-SELECTED AM**, was explained in **Tuning Tricks Challenge SAM**. It negates the necessity of SAM (Synchronous AM) detection.

The S10MK2's tuning indicator LED was replaced with a meter. The unit was powered with two external "D" type alkaline batteries. Earbuds were used for low-power operation. Altering any radio could result in it being damaged. Avoid applying soldering iron heat for long periods of time. **WARNING**: PLEASE wear **EYE PROTECTION** when performing modifications.

Pictured below is the S10MK2 during initial testing (30 gauge wire taps).
The lower pictures show the [air-variable capacitor/toroidal inductor] tuning system.

5. DISCUSSION

The Hellenized ICF-S10MK2 is good for both shortwave program listening and casual DX. Images were a problem due to the low 455 kHz IF and single-conversion circuitry. However, the combination of offset-tuning, narrow filtering, Sony's AGC, and a tuned loop antenna resulted in pleasant sound. There is a plethora of low-priced single-conversion shortwave radios that are not as pleasant to listen to or as fun to operate. This is not so surprising considering Sony has always been a leader in pocket MW radios. The S10MK2 makes a good base for homebrew radio projects.

A new niche, called UltraLight DX, sprung up, years after my 2002 work on using Pocket Radios to MW DX. I have long been a fan of using cheap radios to DX: including pocket radios (SW DX via modification); and one-triode (**Angelodyne**) or one-transistor (**Dee/Mitch-dyne**) designs. Testing the S10MK2 against an R75 opened my eyes to the fact that the voltage induced in an antenna cares not how expensive or how complex the receiver is that converts it to audio. I convert the newer Sony SRF-59 to shortwave in a companion article named **Hellenized SRF-59**.

2.4 Hellenized ICF-S10MK2 Handbook
MW radio hacked to cover SW
©2013

1. INTRODUCTION

This document describes converting the Sony S10MK2 to shortwave. Its companion article is named **Hellenized ICF-S10MK2**. This radio uses Sony's CXA1019S: a bipolar AM/FM radio IC.

2. MATERIALS

This modification requires: a T50-2 toroid, a 100 Ω resistor, a 0.001 µF capacitor, and a variable capacitor (~360 pF) with vernier drive and dial. A 10-turn pot and varactor could be used.
WARNING: Please do modifications while wearing **EYE PROTECTION**.
CAUTION: This modification will permanently alter the radio.

3. OPENING THE S10MK2

Remove the screw inside the battery compartment. Separate the radio in two; there are claws on each side and two on top. Remove the antenna screw and take out the circuit board.

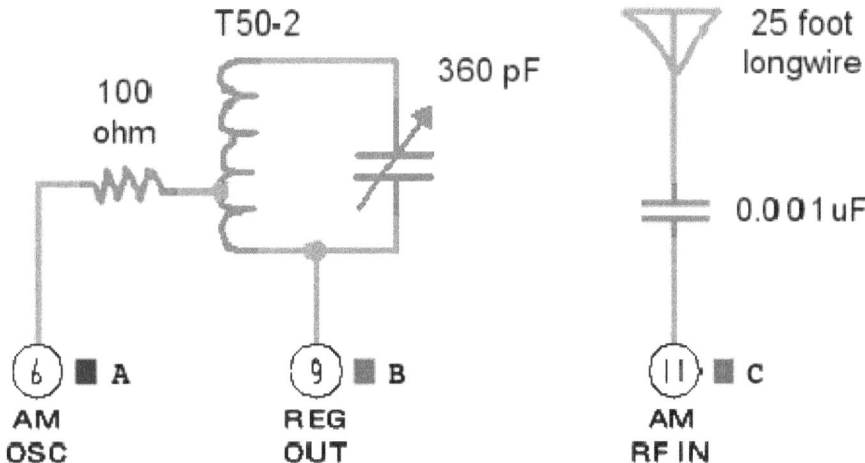

4. ANTENNA MODIFICATION

The AM ferrite antenna was left intact. Shortwave energy, from a 25 foot wire, passes through a 0.001 µF capacitor, and directly into *pin 11*. See "C" above. I opted for a simple design, that is not tuned. The MW tank's inductor prevents SW energy from being bled to ground. Sony typically applies tank input to an emitter, as a low-Z input. Solder to the underside of the chip: the side with the circuit traces and the red LED. A tuned loop can be added for less noise.

5. OSCILLATOR MODIFICATION

The oscillator was altered to receive shortwave. The run to *pin 6*, the local oscillator (LO), was cut using a pocket knife. See the "X" mark, below. The LO tank inductor is an iron powder toroid (µi 10, red); specifically 31 turns of 26 AWG on a T50-2 with a tap at 12 turns for pin 6. The LO tank capacitor is a unit from a signal generator with 2.5:1 vernier drive and dial. The LO alteration needs two taps: *oscillator pin 6 and regulated power pin 9*. Notice that the inductor and capacitor are attached to each other at one end of each component. Sony typically applies tank input to a base, as a medium-Z input. Set the radio to 1605 kHz; this is its lowest capacitance.

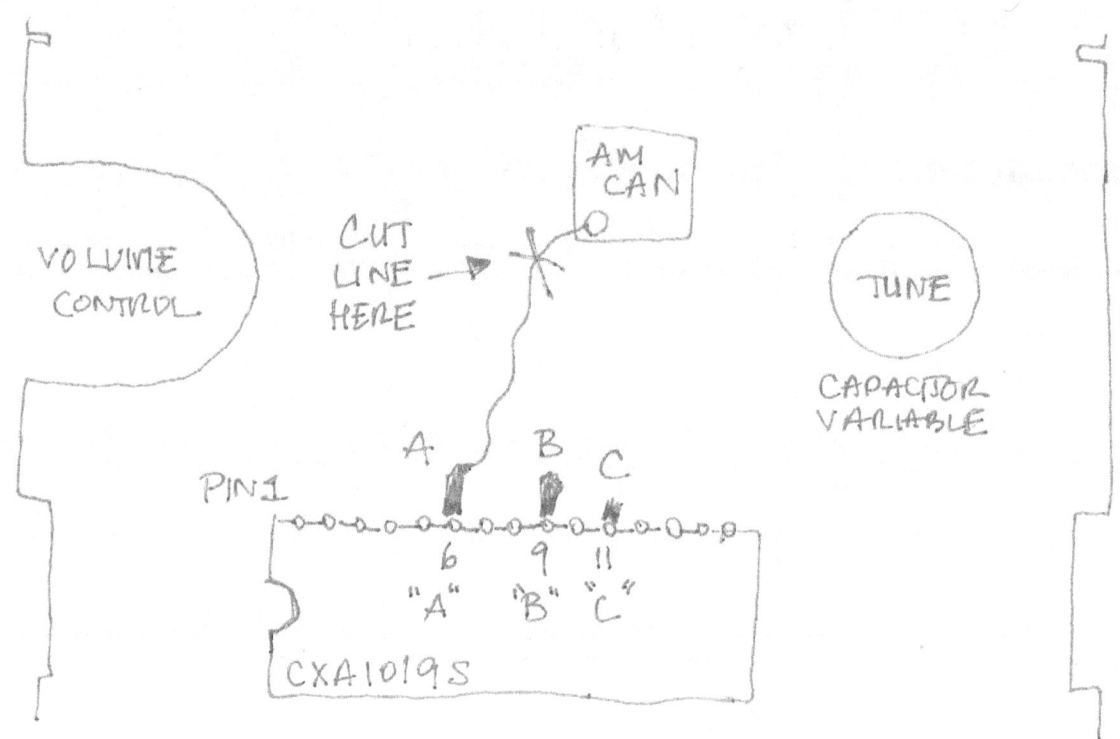

VOLUME CONTROL

CUT LINE HERE →

AM CAN

TUNE

CAPACITOR VARIABLE

A   B   C

PIN 1

6   9   11

"A"   "B" "C"

CXA1019S

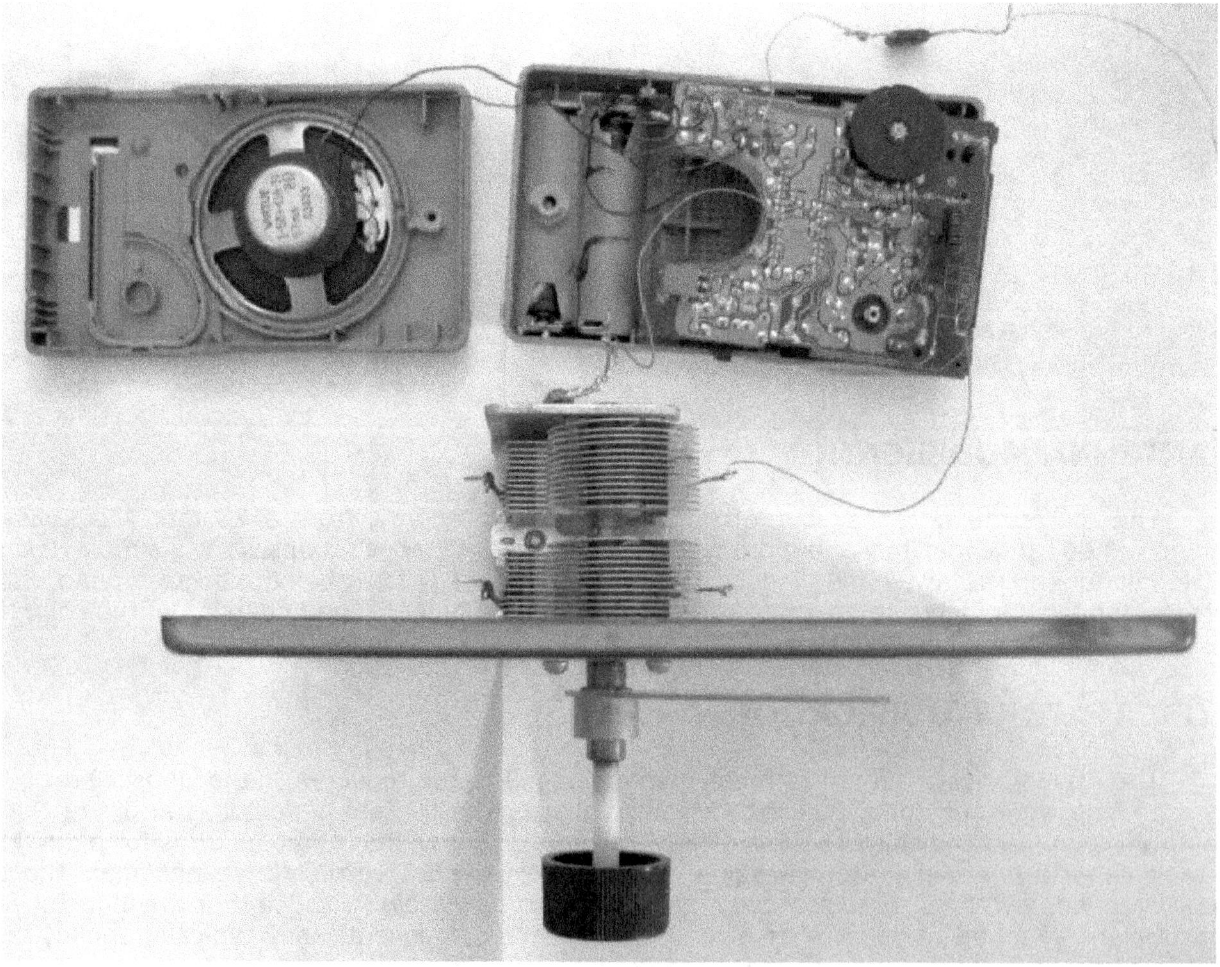

6. DISCUSSION

The Hellenized ICF-S10MK2 is good for shortwave listening and casual DX. Images were a problem due to the low IF and single-conversion circuitry. However, the unit is reminiscent of the old-time superhets. Just set the volume, tune slowly, and listen for stations. For performance I convert the newer Sony SRF-59 to shortwave in a companion article named **Hellenized SRF-59**.

2.5 Hellenized SRF-59
$40 QRP Hartley image rejection low-IF shortwave receiver
©2009

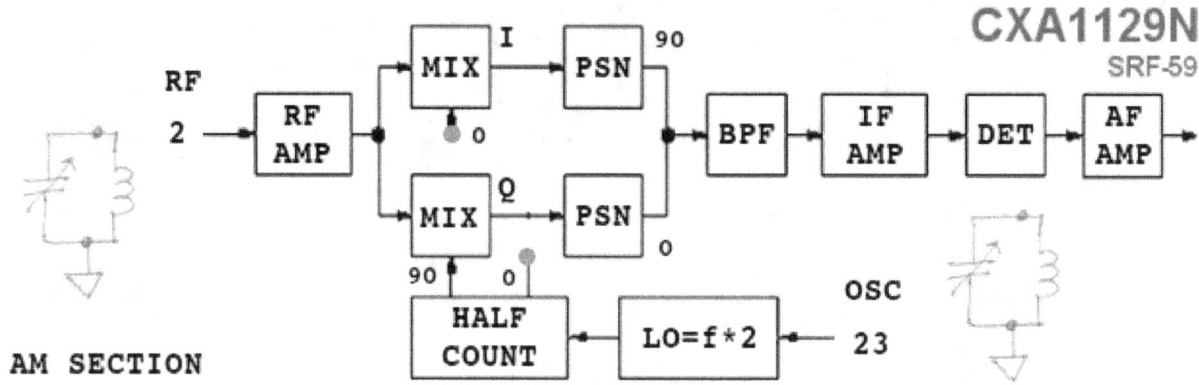

1. INTRODUCTION

My 2003 article named **Sony ICF-S10MK2** proved an $11 radio with a $20 Radio Shack Loop heard, on average, 94% of what an ICOM R75 and Quantum QX Loop could [$780 total]. It found that half the missed stations were faint and half were adjacent to or located at the images of the strongest stations. This effort to quantify using a pocket radio for BCB DX fell on deaf ears. However, years later, a niche, called UltraLight DX was formed to use pocket radios to MW DX. My trusty ICF-S10MK2 was labeled a "turkey" (when used without the RadioShack loop); but it has long since been cooked! My S10MK2's fate was a conversion to shortwave, complete with fifteen element Murata filter. Since the Sony SRF-59 is acclaimed at BCB DX, I decided to study it, alter it, and finally do formal write ups. Read the companion article named **Hellenized ICF-S10MK2**.

My 2006 article named **Tuning Tricks Challenge SAM** explains two tuning "tricks" called **SIDEBAND-SELECTED AM** and **PRECISION ECSS** that negate the need for SAM or Synchronous AM detection. The conclusion was that slow-AGC and sideband suppression reduce carrier dropout related distortion. The article has relevance when studying the proprietary CXA1129N chip within the SRF-59. The article points out using an offset tuned filter to suppress a sideband and diminish fading distortion. Offset tuning, with its resultant increase in fidelity, can be done with an SRF-59.

2. CXA1129N and SRF-59

The CXA1129N is based off a 0.95 Volt bipolar chip designed by *Taiwa Okanobu* in 1992. To eliminate external 455 kHz filters, selectivity would be provided via active RC filters using MIS (Metal Insulator Semiconductor) capacitors on the IC. Their Q (under 20) necessitates usage of a low 55 kHz IF: the problem is the image! The solution is to cancel the image using I/Q [Inphase and Quadrature] double-balanced mixers followed by a PSN (Phase Shift Network). Mixers are fed 90 degree phase shifted local oscillator (LO) signals. The two ~55 kHz PSN's [two 2nd order all-pass] outputs are added and the image is canceled. Image rejection is ~40 dB [~60 dB with a ferrite]. Shortwave reception is possible in the prototype as mixer LO was specified up to 60 MHz. The LO is twice the receive frequency due to the ½ divider. The 55 kHz AM BFP [three 2nd order biquad] has specs of -20 dB at 5 kHz and -40 dB at 10 kHz. *Taiwa*'s test radio showed an AM (6 dB S/N) sensitivity of 3.16 µV [10 dBuV] or S5; S/N of 50 dB; and THD of 0.3%. The chip has three AGC lines: 1) AGC1 at pin 3 at the RF amplifier to limit the input level; 2) AGC2 at pin 9 post-detector as standard IF AGC; and 3) OL AGC at pin 25 post-mixers which is an overload AGC. The SRF-59 uses a fixed resistor off GAIN ADJ at pin 26. The prototype tunes mixer gain via a potentiometer, which maximizes image rejection. Tuning indication [pin 22 is TUN IND] and stereo FM indication [pin 15 is ST IND] can be added to a SRF-59. Sony's schematic contains an error: D2, one of the FM antenna input protection diodes, needs to be flipped so that its cathode is attached to ground.

3. QRP ARCHITECTURE

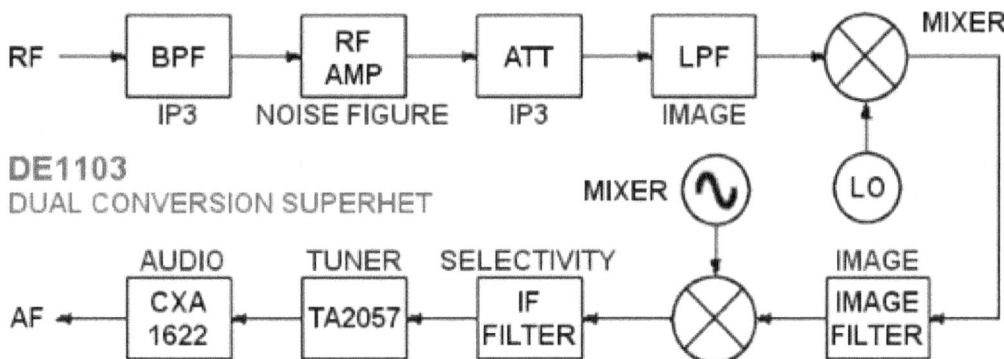

In Phil's SW Radio Picks I recommended several portables including the DE1102, DE1103, and KA2100. These double-conversion PLL-synthesized portables have good image rejection and selectivity. They are also power hogs that supply an AM/FM tuner, CPU, PLL, and audio chips, as well as external mixers and LCD displays. The DE1103 using earphones and not backlit consumes 71.3 mA at 6 Volts or 428 mW of power. Four Duracell "AA" alkaline batteries will last ~40 hours. By comparison the low-power (QRP) SRF-59 uses 15.8 mA at 1.5 Volts or 24 mW of power. One Duracell "D" type battery *may* power the radio for 950 hours or 2.6 hours daily for an entire year.

Zero Intermediate Frequency (Zero-IF or Direct Conversion) receivers have no image and are QRP. Unfortunately Zero-IF suffers from DC offset (parasitic LO coupling to the antenna, LNA, and mixer), flicker (1/f) noise, LO antenna leak, and even order distortion. Zero-IF receivers are also notorious for problems with dynamic range, AC hum, microphonics, and weak audio output.

4. HARTLEY IMAGE REJECTION

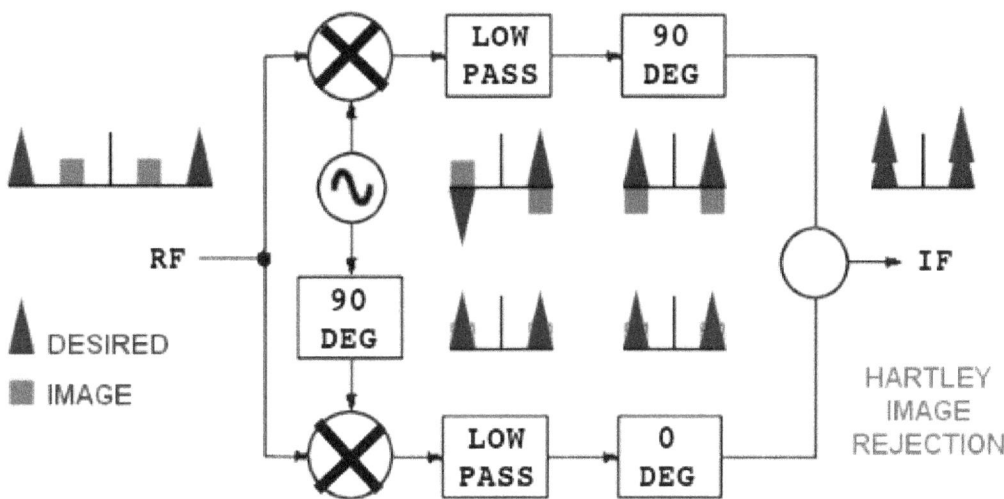

Sony's engineers knew just how to get good overall performance under low-power. They combined a low-IF architecture, which does not suffer from much DC offset or flicker noise, with Hartley image rejection. Although phase and amplitude mismatches limit this image rejection to 40 dB the rest can be made up with a tuned antenna. This is a perfect symbiosis: low-IF suffers from images which Hartley (1928) architecture addresses [Weaver produces secondary images].

Unlike others, Sony has the engineering power to make their own *low external parts* chip. The SRF-59 contains no ceramic IF filters; and no power-using tuning LED or DSP IF. A separate LA4537M headphone power amplifier was used to prevent the need for balanced amplifiers inside the CXA1129N. This avoids feedback of audio, and its harmonics, into the RF amplifier or mixers.

5. MODIFICATIONS

The conversion of an analog tuned radio from BCB to SW must address changing the local oscillator (LO) and antenna. The LO tank inductor may be replaced with a red type-2 iron powder toroid. The LO tank variable capacitor may be swapped for an air-variable capacitor with vernier drive or varactor diodes with a 10-turn potentiometer. The BCB ferrite antenna must be replaced. Untuned antennas can be used; but tuned loops reduce various mixer artifacts and can be rotated to attenuate local noise. Magnetic loops can be made using a tank with long connections between the capacitor and toroidal inductor [copper tubing can be used]. Statistically 4.7 MHz to 10.0 MHz covers ~**89.6%** of all evening English broadcasts beamed to America [by hours of programming]. Analog oscillators with tuned inputs have no CPU or synthesizer noise and low stray mixer energy.

Advanced radio alterations include changing the external AGC capacitor: reducing it during band scanning and increasing it (ex. ~220 uF) for listening [switch selects]. The latter can reduce selective fading distortion. On the audio end, a speaker, audio amplifier (TDA7052 or LM386), and audio filtration (passive **RC** or two-pole low-pass op-amp) make good additions. Comfortable $13 Koss Sparkplug earbuds are my favorite [112 dB SPL per mW]. Please note that modifying radios might result in their damage. To learn more about soldering to IC pins see **Phil's Soldering 101**.
WARNING: PLEASE wear **EYE PROTECTION** when performing modifications.

6. HELLENIZED SRF-59

I do NOT recommend modifying your SRF-59: the radio is easily ruined. This description is for amusement purposes. The tank's variable capacitor was rendered inoperable by removing two capacitors. The tank's inductor was removed. Four taps were created: *oscillator, antenna, ground,* and *power*. An old Heathkit Signal Generator was used for its tuning dial, vernier, and air-variable capacitor. An Amidon toroid was used for an inductor. The antenna pictured is a tuned loop: input via the ferrite's tap feeds 1.5 V to pin 2. The ferrite at SW acts as resistance for the chip's internal RF amplifier. Soldering was difficult (easily lifted lines). Superior circuits are being tested. The top pictures illustrate the size of the SRF-59 and its internals; bottom pictures are of the SW test unit.

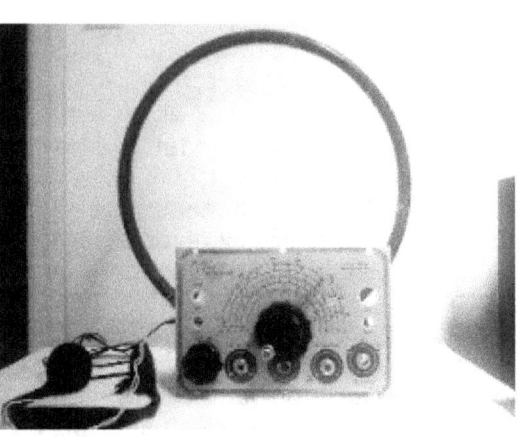
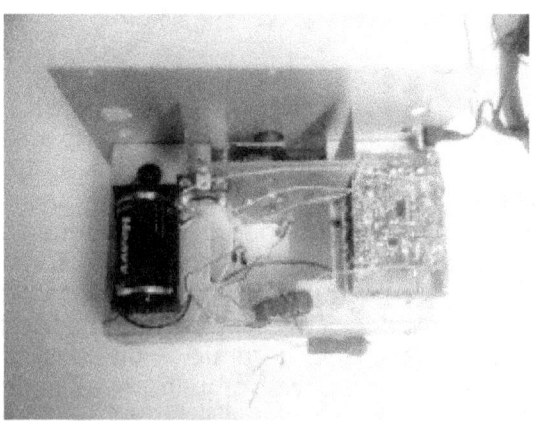

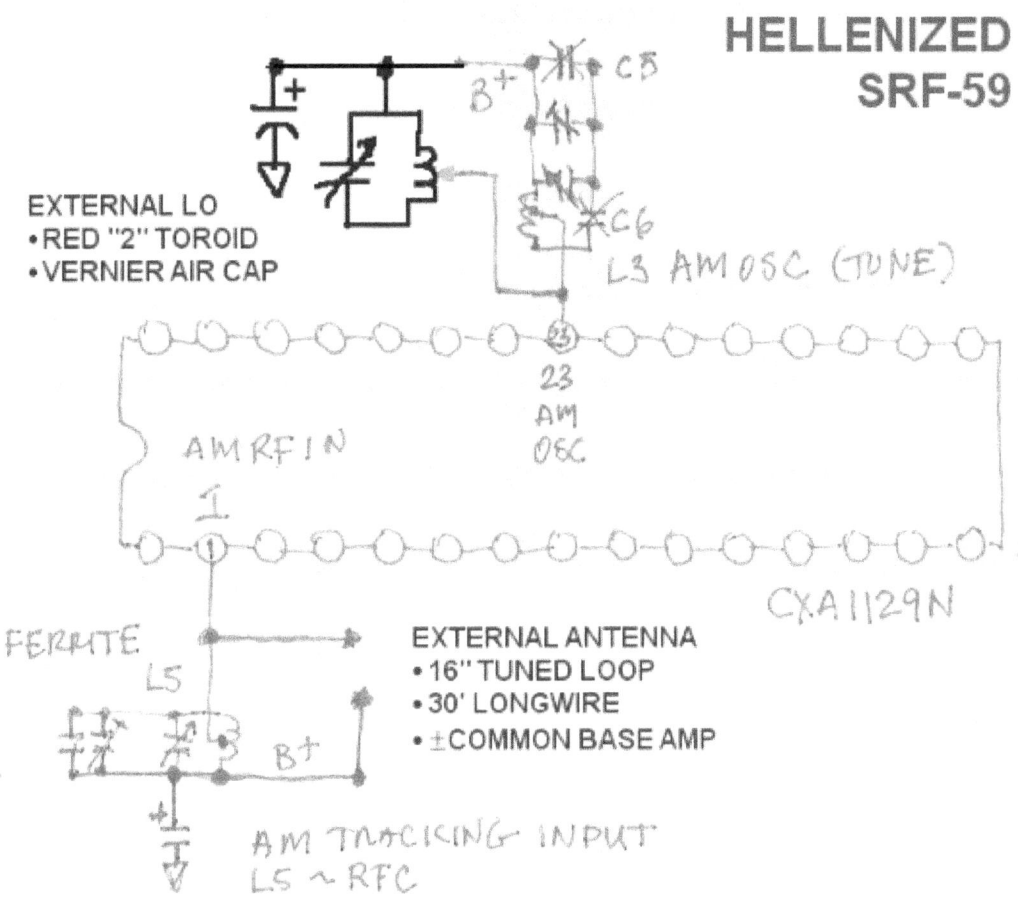

HELLENIZED SRF-59

EXTERNAL LO
• RED "2" TOROID
• VERNIER AIR CAP

L3 AM OSC (TUNE)

23 AM OSC

AM RF IN 1

CXA1129N

FERRITE L5

EXTERNAL ANTENNA
• 16" TUNED LOOP
• 30' LONGWIRE
• ±COMMON BASE AMP

AM TRACKING INPUT
L5 ~ RFC

7. DISCUSSION

The Hellenized SRF-59 rivals $4000 receivers in architectural complexity. Shortwave is a niche market: engineering muscle is concentrated in the lucrative BCB/FM consumer market. As hackers we can convert MW designs to shortwave and get a taste of the new technology without the high cost. Degen is using Silicon Lab's Si4734 AM/FM/SW/LW receiver IC with DSP IF in their $70 DE1123. Sony is using their CXA1376AS receiver IC with SAM in their $146 ICF-SW7600GR. Choices are obviously limited in the low-cost high-tech shortwave marketplace. The swapping of dual-conversion and ceramic filters for I/Q image-rejecting mixers and RC filters has merit when considering QRP usage. The Hellenized SRF-59 has a unique sound: warm, clean, and bass filled.

The Hellenized SRF-59 does great at program listening with a tuned loop: images were rare. The audio is tight, with fidelity being increased by offset-tuning. Sony did well with the AGC. DX tests and circuit optimization are ongoing. A SRF-59 displaces 9.7 cubic inches. The Hellenized SRF-59 represents a lot of shortwave listening "***bang for the buck***". You can now build a QRP Hartley image rejection low-IF shortwave radio with tuned loop antenna for $40. QRP amateurs, AM (LW/MW/SW) aficionados, and FM DXers may find use in modifying Sony's SRF-59.

Want to DX with less radio? My 2008 article named **Angelodyne** describes a new vacuum tube detector called a *regenerative plate detector*. Hear the world using a single frame-grid triode, modern 16-ohm earbuds, and non-lethal plate voltages.

2.6 Hellenized SRF-59 Handbook
QRP Image-Rejecting Low-IF Shortwave Receiver
©2011

1. SONY'S UNIQUE SRF-59

This document describes converting the Sony SRF-59 to shortwave. Its companion article is named **Hellenized SRF-59**. The Sony contains a robust proprietary bipolar junction transistor chip: the CXA1129N. Selectivity is achieved via active 55-kHz RC intermediate frequency filters; image rejection consists of adding the output of phase shift networks that follow two I/Q mixers. The SRF-59 is a high-tech alternative to using dual-conversion and ceramic IF filters. This analog design is both low-noise and low-power. The SRF-59 can run 180 hours on a single "AA" battery.

Most quality shortwave receivers are digital to assist tuning. However, their topology also contains a noise-producing microprocessor, liquid crystal display, and synthesizer. Band scanning for strong stations is much easier on an analog set. Double conversion means more in-line mixers and more noise and power. The SRF-59's 55-kHz filters sound better than many 450 and 455-kHz filters. The Hellenized SRF-59's sound is warm, clean, and rich. It is as good as any radio I have heard; including the Drake-designed Satellit 800 and Eton E1. Its sensitivity and selectivity permits both long-distance DXing and enjoyable program listening. The SRF-59 costs ~$20, shipped.

2. MATERIALS

The Hellenized SRF-59 modification requires five parts: two T50-2 toroids, a 100 Ω resistor, a 0.22 µF capacitor, and a variable capacitor. Ideally a chassis will house an air-variable capacitor (ex. 360 pF) with a large tuning knob, vernier drive, and dial indicator. The iron powder toroids can be bought from Amidon. The stout Sony SRF-39 and SRF-49 (available used) can also be modified.

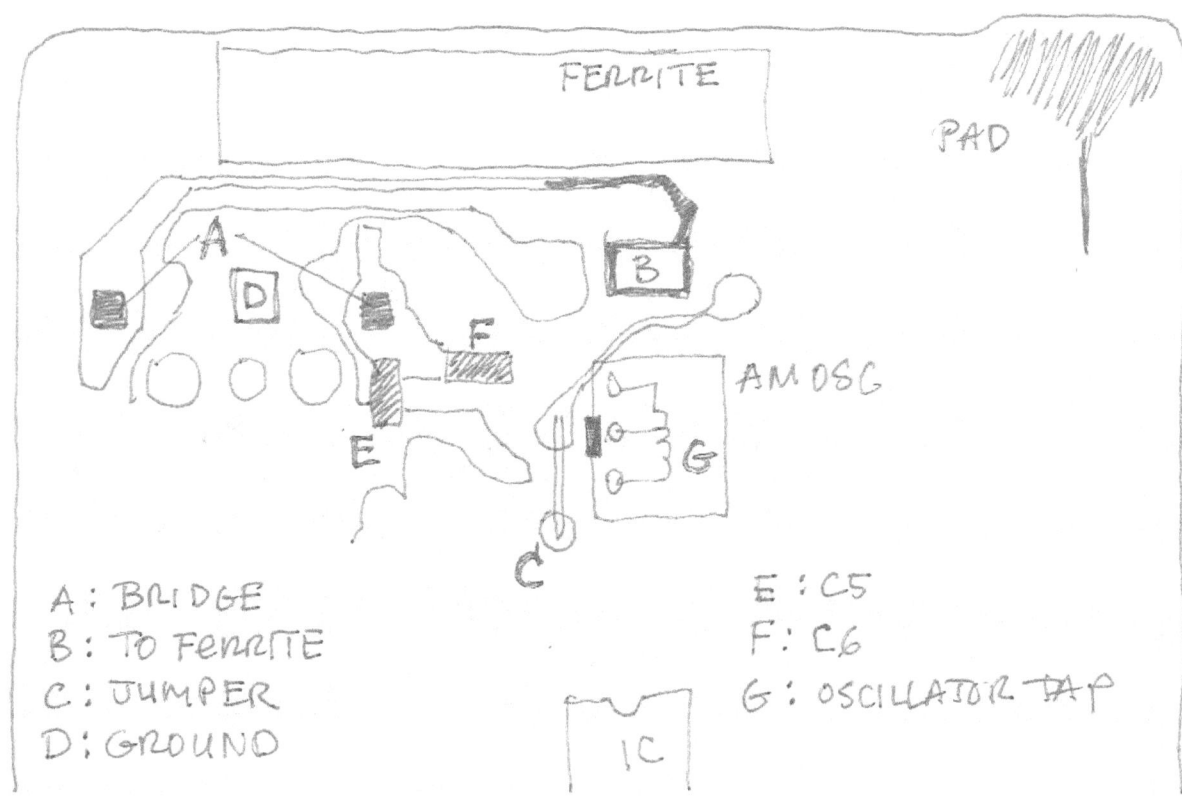

A: BRIDGE
B: TO FERRITE
C: JUMPER
D: GROUND

E: C5
F: C6
G: OSCILLATOR TAP

3. SHORTWAVE FREQUENCIES

The hottest shortwave bands during the nighttime are 49M, 41M, and 31M; or 5.7 to 10.0 MHz. Daytime hot spots include 25M, 22M, 19M, and 16M; or 11.5 to 17.9 MHz. This modification will center on receiving ~5.7 MHz to 17.9 MHz to include the majority of shortwave programming.
WARNING: Please do modifications while wearing **EYE PROTECTION**.
CAUTION: Modifying radios can easily result in their damage.

4. OPENING THE SRF-59

To open the SRF-59 simply remove the two screws from the back, lift the circuit board at the battery end, and gently pull it rearward so that the headphone jack clears its retaining circle.

5. ANTENNA MODIFICATION

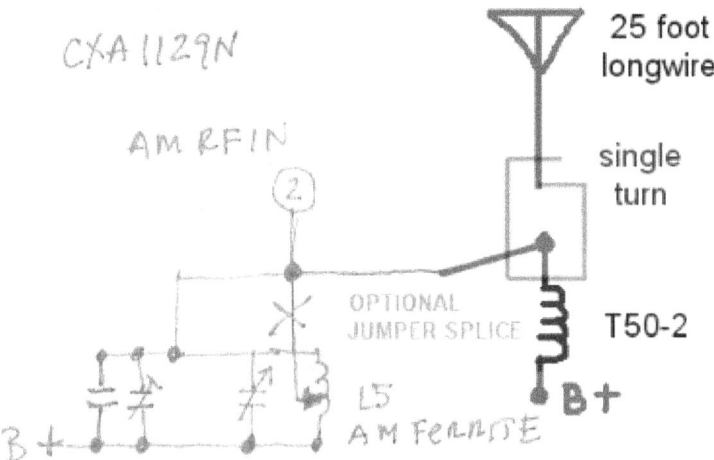

The AM ferrite rod antenna system must be altered for shortwave reception. Sony uses a tuned input. Their engineers typically apply antenna tank input to a transistor's emitter, making them a low-Z input. This invites tapping down on the tank for current. SRF-59 sensitivity is ample without an amplifier: ~1.6 µV or S4. The T50-2 toroid could be omitted (at shortwave the ferrite appears as resistance); but it is critical to high performance to reduce mixer energy. We prevent overload by using a tuned input consisting of 25 turns on a T50-2 toroid, or 3.1 µH. The SRF-59 tuning wheel is used as an antenna tuner. I left the ferrite rod intact: MW energy is dumped and inductance does not add in parallel. You can experiment using a (tuned) loop instead of the toroid.

Although B+ (1.5 V) is applied to the chip, it is key to note that B+ at radio frequencies is at ground due to three decoupling capacitors. The incoming antenna wire is looped a single turn around the antenna's tuning toroid and then back on itself to form a gimmick capacitor (this can reduce FM noise). The design prevents DC from entering AMRF_IN and prevents overload (which easily occurs with antennas hooked more directly to the toroid). With overload there will be hiss instead of any radio stations. I found the Q or quality factor of the antenna tuner to be excellent. Tuning is sharp and this prevents stray RF from entering the chip. One caveat: with the antenna detuned even strong stations will sound poor. See below for details on properly tuning the radio.

Look closely at the BOARD layout on the previous page. The B, C, and D are needed for the antenna system changes. Study the side of the board with the two chips and notice how the lines correspond. The point "B" is a tap to a big solder blob on the tuning capacitor. This provides B+ power beyond the power switch. It is the bottom of the tank. The point "B" is a tap to another solder blob and is the top of the tank. The line at point "C" is the tap going into the chip's AMRF_IN (pin 2). The line is a jumper on the component side: it can be cut to bypass the ferrite. Avoid damaging the jumper because the chip's pins are spaced too close to solder on. The blocks at "A" were connected via a wire. This mod will add the oscillator's variable capacitor (CV1 ¾) to the antenna tuner. A photograph below shows how to protect the chip's 4 taps using a solder lug.

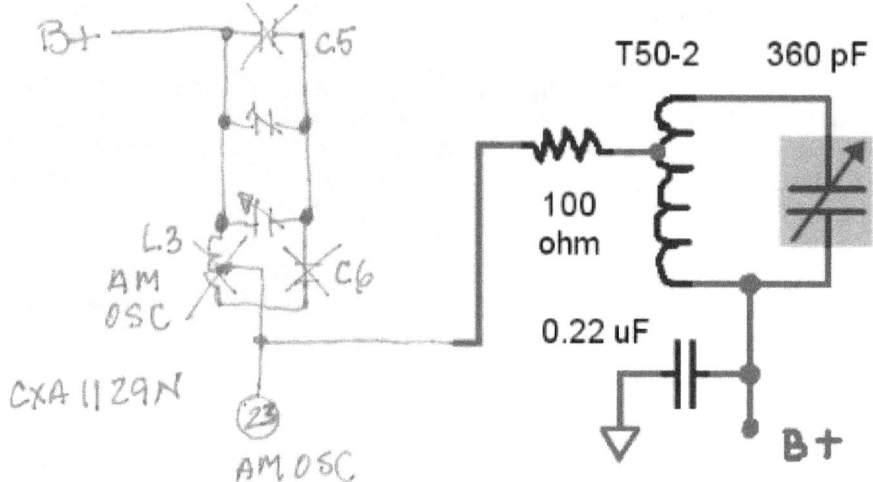

6. OSCILLATOR MODIFICATION

The local oscillator (LO) must be altered for shortwave reception. It is essential to realize that the LO is twice the receive frequency due to the CXA1129N's internal ½ divider. The higher tuning range, above 15 MHz, would ideally use a different toroid. A T50-2 toroid is resonant from 1 MHz to 30 MHz, meaning it can tune the SRF-59 up to ~15 MHz. Sony engineers typically apply oscillator tank input to a transistor's base, making this a high-Z input. Unlike a FET, bipolar bases respond to current, not voltage. It was necessary to use a 100 ohm resistor in the circuit. At SW frequencies, 100 Ω has the impedance of a 150 pF cap; but allows DC to enter AM_OSC (pin 23). Without resistance, tuning was not effective. I used 12 turns on a T50-2, or 706 nH. The tap was placed 5 turns from the top of the tank. You can experiment with turns, tap, and omission of the resistor. I used a 0.22 uF capacitor (value is non-critical) to bypass any stray RF at B+ to ground.

I used my reliable Heathkit Signal Generator with its tuning dial, vernier, and air-variable capacitor of 360 pF. Using a smaller capacitor or putting one in series will make tuning the radio easier but cover less range. Different coils can be used or a larger type "2" toroid, bus wire, and alligator clips. Another viable tuning option is using varactor diodes and a 10-turn potentiometer.

WARNING: this section will alter your radio and could result in it being destroyed. The AM oscillator must be removed. It is on the component side of the board: a silver metal box with red circle. See the BOARD layout. This canister must be de-soldered. It is crucial not to apply heat for extended periods: the board is fragile and lines can easily become lifted. The internal capacitor is rendered inoperable by removing two capacitors: C5 and C6. See points "E" and "F" on the BOARD layout. You can de-solder them or grasp them with pliers and torque them off. The rectangle near "G" on the BOARD layout indicates the location of the pad where the oscillator tap is soldered.

7. AGC MODIFICATION

The stock SRF-59 has an excellent automatic gain control (AGC). It is beneficial to utilize fast-AGC during band scanning and slow-AGC during listening. The AGC is slowed by addition of capacitance in parallel to the existing capacitor. The CXA1129N's three AGC pins can be traced: then capacitance added. The RF-amplifier AGC1 is pin 3; the post-detector AGC2 is pin 9; and a post-mixer OL_AGC is pin 25. Testing suggests that adding 200 µF to pin 25 might be beneficial.

8. OPERATING THE MODIFIED SRF-59

To use the Hellenized SRF-59: apply power, set the volume, and set the band to AM. The stock headset may be replaced with earbuds. Tune the radio to a vacant spot within the band of interest. Slowly move the stock tuning wheel until the hissing sound is maximized. The antenna cannot be tuned while sitting on a station (this will simply elicit AGC action). Now tune the radio very slowly and listen for stations. The antenna tuning can be re-adjusted. Offset tuning (tuning to one side of a station) can be utilized to increase fidelity or help with selective fading distortion.

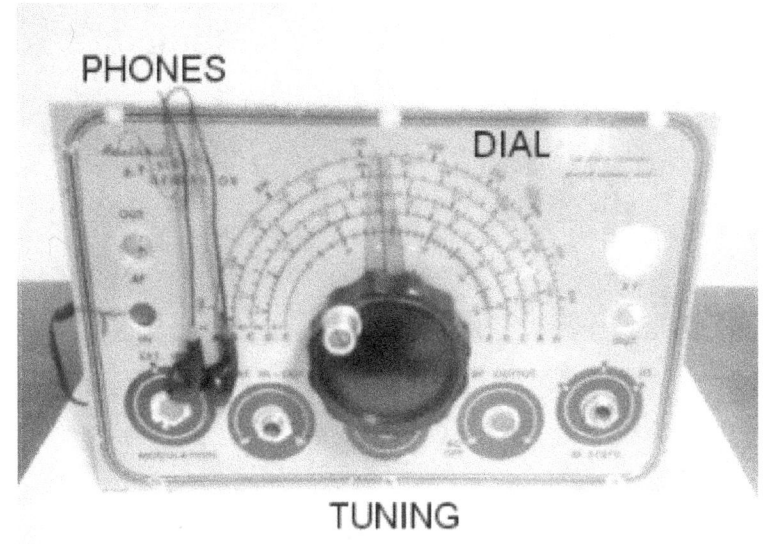

PHONES

DIAL

TUNING

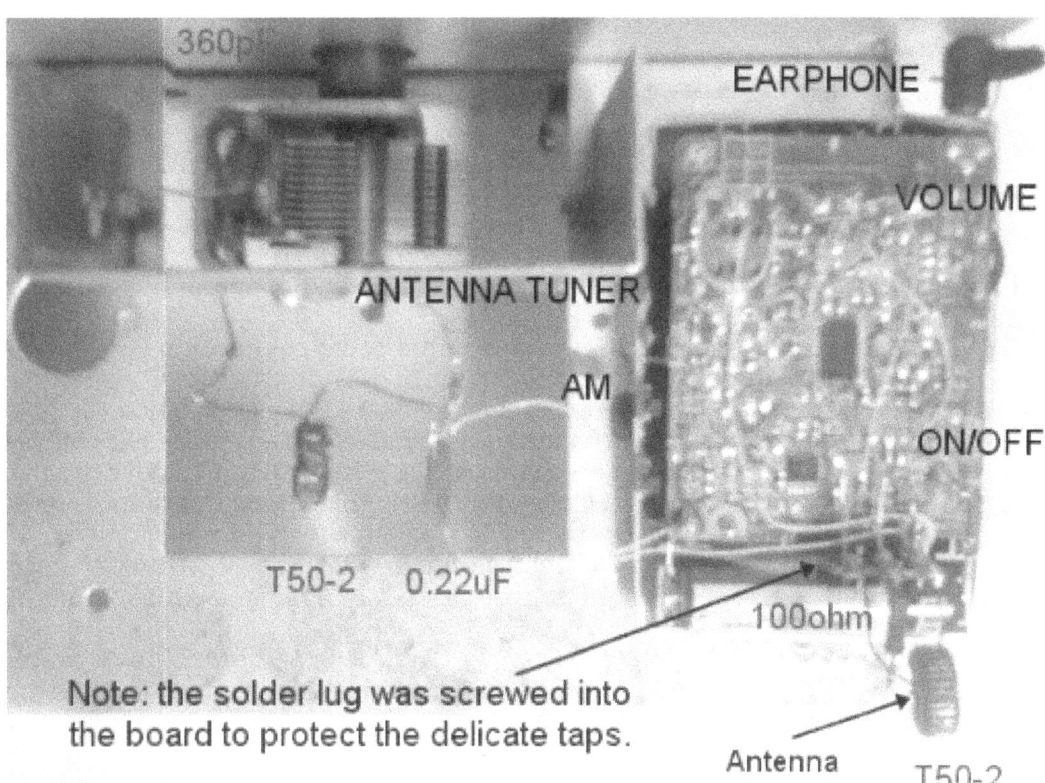

360pf

EARPHONE

VOLUME

ANTENNA TUNER

AM

ON/OFF

T50-2   0.22uF

100ohm

Note: the solder lug was screwed into
the board to protect the delicate taps.

Antenna   T50-2

9. DISCUSSION

In **Phil's SW Radio Picks** I recommend some portables including the KA2100. Although the Hellenized SRF-59 is not as sensitive it can essentially hear anything the KA2100 could. The modified SRF-59 makes a solid shortwave DXer. The radio shines during program listening: the 55-kHz filters have a unique sound, Sony's detector is first-rate, and listening hour after hour is not tiring. There is also ample room for improvement of this base design by other radio hackers.

2.7 Hellenized CXA1600P
Sony's Secret $7 AM Only Radio IC
©2014

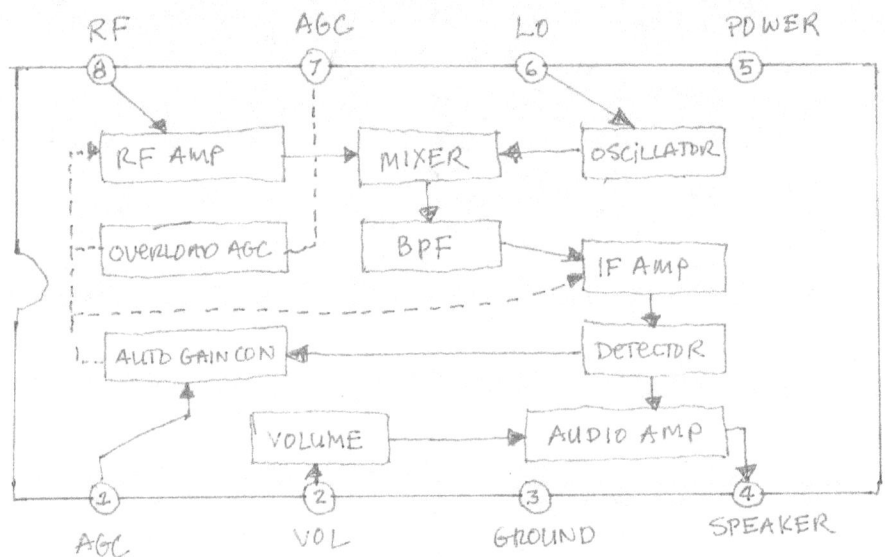

Sony describes their CXA1600P bipolar IC as an: 8-pin (DIP) Single-chip AM Radio with Built-in Power Amplifier (100 mW). Supply is 1.8 to 4.5 Volts. Current was measured at 6.7 mA (at 3V, with audio). Selectivity at 55 kHz (low-IF) is: -17 dB @ -10 kHz and -23 dB @ +10 kHz. The spec sheet shows the equation: $f_{if} = 1/2\ f_{osc} - f_{IN}$. The block diagram (see above) shows a single oscillator and mixer, between the radio frequency (RF) amplifier and band-pass filter (BPF).

Sony's CXA1129N (used in the SRF-59) is based on a 1992 chip design by Taiwa Okanobu, Hitoshi Tomiyama, and Hiroshi Arimoto. Also, briefly described, is an "AM Only Radio IC" with pins identical to the CXA1600P (with a balanced audio amplifier). The significance is that the AM section of an SRF-59 was built into a superb shortwave radio (See Hellenized SRF-59). However, many will be unable to reproduce my work and the CXA1129N is too tiny to solder on. In contrast, the CXA1600P can use a DIP socket. It is capable of 150 kHz to 30 MHz operation. My prototype exhibits few images. The equation above points to the oscillator feeding a half-divider. This likely feeds two double-balanced mixers (I/Q) that enter two phase-shift networks (that are summed to cancel images). This is what is actually located between the CXA1600P's RF amplifier and BPF.

The CXA1600P is ideal for homebrew DX on LW, MW, or SW. There is no noise-producing: PLL, MCU, or LCD. Hartley image rejection (~40 dB) replaces dual conversion. And internal low-IF, 55-kHz, RC filters replace external IF transformers and ceramic IF filters. Unlike the CXA1129N, there is no tuning indicator. Power is minimal. The CXA1129N (mated to an audio amplifier: LA4537M) uses 24 mW. The CXA1600P uses 20 mW. For comparison: the DE1103 uses 428 mW.

Sony makes two main analog shortwave radios. The $67 ICF-SW11 that is single conversion. And the $148 ICF-SW23 that is double conversion with a single JFET first mixer.

To build your own CXA1600P MW "Pocket Radio", follow the data sheet. The Hellenized CXA1600P builds the radio for shortwave. The aim is 5.7 MHz to 17.9 MHz, covering the active night and day shortwave bands. Sensitivity is 1.6 µV. This paper seeks to **reduce the learning curve.** Make all connections carrying RF as short as possible. It is now possible to build a sturdy and easy-tuning MW, LW, or SW radio with SRF-59 performance. This is just a prototype (subject to change). At the time of building this project there were no references; however, I was beat to the punch by this Russian link: *http://radiocon-net.narod.ru/page24.htm* (now 404).

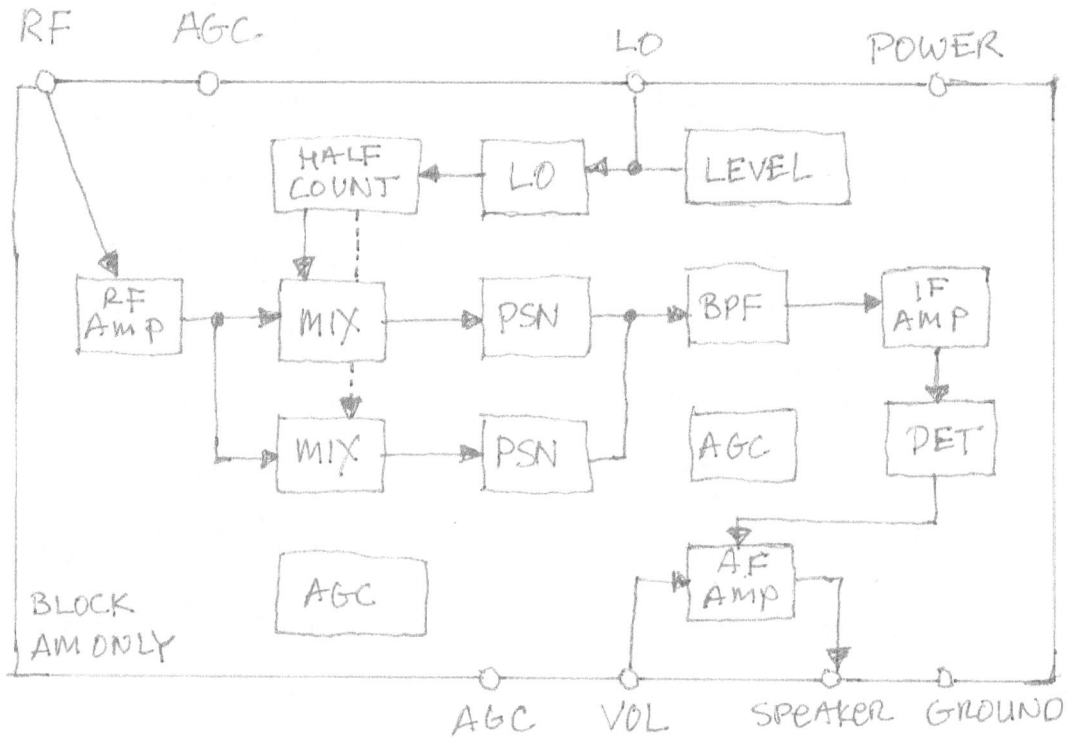

The block diagram of Sony's "AM Only Radio IC".
WARNING: PLEASE wear EYE PROTECTION when building your radio.

PIN 1: AGC (Automatic Gain Control)

Shown is a 22 µF capacitor. I use 100 µF for slow-AGC (listening).

PIN 2: Volume

Shown is a 100k ohm pot. I use a fixed 10k or 22k ohm resistor: not ideal but workable. Grounding the pin causes the lowest volume; whereas, 100k ohms produces the highest volume.

PIN 3: Ground

Use solid bus wire for a physically stable ground.

PIN 4: Audio Output **CRITICAL**

Avoid driving an 8-ohm speaker. I use headphones at 16 ohms (two 32 ohm in parallel; mono). I use a single 0.22 µF bypass capacitor and an RF choke.

PIN 5: Power

I use 2 "D" cells. Shown is a 220 µF power capacitor: I use a 2200 µF.

PIN 6: Oscillator **CRITICAL**

An old-school dial cord tuning system would work well. I use a 360 pF air-variable capacitor with vernier drive, dial indicator, and a large knob. The tank inductor is a red, Amidon, iron powder, T50-2 toroid. I use 12 turns with a 2 turn tap from the top of the tank, feeding pin 6. This is subject to change. The oscillator is set to twice the receiver frequency plus 110 kHz. A 0.22 µF capacitor (value is not critical) is used on the tank base to bypass any stray RF to ground. A varactor tuning diode with a 10-turn pot and/or multiple tanks (selectable bands) could be used.

PIN 7: Overload AGC

Shown is a 4.7 uF capacitor and 2 silicon diodes (in series). I use a 10 uF.

PIN 8: Antenna **VERY CRITICAL**

The CXA1600P is very sensitive. I use an RF choke (30 turns on a T50-2) from power to pin 8 (to provide 3 V). The antenna is 3 turns from the pin end, attached via an RF capacitor. Alternatively, the antenna is wrapped around the pin end (post) to create a <1 pF gimmick capacitor. My antenna is 25 feet of wire. Ideally, a tuned input should be used to reduce unwanted mixer energy. The antenna would be attached to a tank top: the tank bottom is attached to RF ground (power is at RF ground). A variometer (two variably meshed inductors) might be ideal for antenna input. If there is loud hissing, the radio is overloaded.

Hellenized CXA1600P
Prototype One

2.8 Tuning Tricks Challenge SAM
Dispelling Shortwave Myths
©2006

1. INTRODUCTION

Recently *Dallas Lankford* stated: "*The more I study and use AM synchronous detectors the more I am mystified as to why they are so highly acclaimed*". It is possible synchronous AM (SAM) detectors are praised because: 1) tuning tricks are not being employed, and 2) most SW receivers do not have accurate or stable enough frequency synthesis for optimum SSB reception. Most SAM units do not live up to that inside the legendary R8B. Note: DSB (double-sideband full-carrier) SW signals and SSB (single-sideband suppressed-carrier) ham signals are both amplitude modulated.

2. Sideband-Selected AM Trick R75 COOKBOOK "AM DETUNING TRICK" 2002

A. Select AM-mode, narrow filter (ex. 4-kHz), and slow-AGC.
B. Offset-tune by plus (USB) or minus (LSB) nearly half the filter's bandwidth.

This trick minimizes selective fading distortion and allows selection of the cleaner sideband (determined by SSB). With a 4-kHz filter the BBC at 5975 kHz is USB tuned at 5977 kHz and LSB tuned at 5973 kHz. Audio (fidelity) brightens most near half the filter's bandwidth. PBT can attain similar results. Ceramic filters often work better than mechanicals due to their wider skirts. Under *normal* selective fading this trick works well and is feasible on low-cost portables. Look for radios with double-conversion, PLL-synthesis, 1-kHz steps, and a narrow filter (ex. DE1102 or DE1103).

Many ways exist to select one sideband of a DSB signal. The filter method is often done automatically in SSB-mode. The trick above uses a filter in AM-mode. Another method is audio phasing using op-amps (operational amplifiers; ex. E1) or PSN (phase shift networks). PSN allows more adjacent sideband rejection (over 70 dB versus ~30 dB) and less distortion (see *Polyphase Network Calculation using a Vector Analysis Method*) than op-amps. An added method is software.

3. Precision ECSS Trick R75 COOKBOOK "ECSS FINE TUNING TRICK" 2002

A. Select SSB-mode, wide filter (ex. 6-kHz), and slow-AGC.
B. Fine-tune in 1-Hz steps until flutter stops.
C. Engage the SSB filter (ex. 2.5-kHz) for listening.

This trick quickly and precisely tunes the BFO to within 2-Hz or less of a station's carrier. This is ECSS: tuning DSBc signals as SSB. Flutter occurs because: 1) the wide filter passes both sidebands, and 2) a discrepancy exists between the BFO and carrier (the resultant tone is too low to hear). A single tone (ex. 1000 Hz; 3 Hz off) is heard as two sideband tones (ex. 997 Hz, 1003 Hz). This disparity allows effortless BFO fine-tuning (see JND below). Small tuning steps (1-Hz) and high stability (±1 ppm) are needed for precise ECSS. Stations are quite stable. Under *heavy* selective fading ECSS is *superior* to SAM. ECCS requires no lock and works on faint signals (DX).

The acronym JND stands for "Just Noticeable Difference". The human ear has a JND of 1.5 Hz at 500 Hz. Given a 500 Hz tone, a tone of 501.5 Hz or greater is detectable as being different. JND varies with frequency: JND is 2.9 Hz at 1000 Hz, 5.8 Hz at 2000 Hz, and 8.7 Hz at 3000 Hz. The ear acts as a mechanical spectrum analyzer; sensing amplitudes for each frequency. Humans cannot hear phase differences between the different components of a tone: Ohm's law of hearing.

4. CARRIER DROPOUT DISTORTION

A 50,000 Watt DSB transmission at 30% modulation outputs 1,125 Watts of power into each sideband. Total power is 52,250 Watts and together both sidebands, which contain *all* the audio information, receive only 4.3% of that power. Carrier power, which is held *constant*, will receive the other 95.7%. DSB "wastes" power in order to permit low-cost envelope detection.

Diode detection of DSB is forgiving of unstable local oscillators (LO). During LO frequency shifts, the sidebands move in step with the carrier: the carrier, akin to being a BFO, is essentially perfectly "synchronized". For lower distortion detectors look up precision half-wave rectifiers (see *An Improved Precision Full-wave AM Detector* and *Low Distortion/High Dynamic Range Detector*).

Selective fading causes sideband attenuation and carrier dropouts. The later can make SW listening unpleasant. During carrier reduction, signals appear over-modulated. With regular diode detection, total harmonic distortion (THD) increases with modulation index. Although with normal modulation, diodes are capable of fairly *low* distortion: 0.14% THD at 30% modulation index. The best SAM distortion value is 0.40% for the R8B. Other SAM distortion values are: 8.20% for the WJ-8711A, 2.6% for the RX-340, 2.0% for the 7030, 2.4% for the Sat800, and 2.6% for the E1.

5. ILLUSTRATIONS

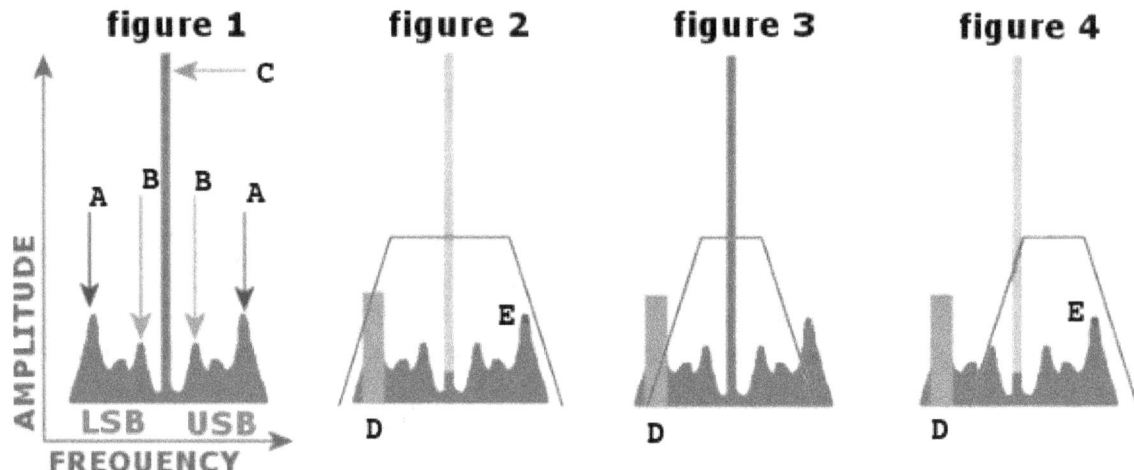

Figure 1: High amplitude (see "C") carrier energy and positions of LSB and USB audio. Lower tones (1000 Hz; see "B") are close to the carrier; higher tones (3000 Hz; see "A") are father away. Sidebands can be seen dancing in unison. **Figure 2:** Wide filter (6-kHz) passes the offending tones (see bar at "D") in AM-mode. Carrier dropout is depicted by a shaded carrier dipping into the noise floor. **Figure 3:** Centered narrow filter (4-kHz) with low fidelity (2000 Hz @ -6 dB) passes offending tones (see "D"). **Figure 4:** Upward (USB) offset-tuning of the narrow (4-kHz) filter (or applying PBT) attenuates the offending tones (see "D") and increases fidelity (4000 Hz @ -6 dB). This illustrates selecting the cleaner sideband of a DSB signal via the filter method in AM-mode.

Consider carrier dropouts in **figure 2** and **figure 4**. Imagine mixing between the USB peak (see "E") and frequencies farthest to the left during carrier loss. The centered 6-kHz wide filter allows 3-kHz fidelity but noises up to 6-kHz can be emitted during a carrier dropout. However, the offset-tuned 4-kHz wide filter allows 4-kHz of fidelity but noises up to *only* 4-kHz can be emitted. Audio filtering, which can reduce carrier dropout distortion, is less vital when using offset-tuning. Improper usage of a fast-AGC (normally used for scanning or CW) further complicates the carrier dropout problem by activation of the radio's AGC system (carriers contain most of the RF energy).

6. RADIO DESIGN for TUNING TRICKS

A quality receiver will utilize double-conversion for image rejection (up-conversion), gain distribution, and filter shape (down-converting to 455-kHz). Two filter bandwidths (7-kHz; 4-kHz) could be mated to a ±3.5 kHz PBT, and a matched-LC phase shift network. Add toggle controls for **SIDEBAND-SELECTED AM** (local oscillator shifter) and **PRECISION ECSS** (shutting off the PSN). The unit should have 1-Hz tuning steps, ±1 ppm stability, and a slow-AGC (two second plus release time).

Examples close to the radio described above are the R75 (missing audio phasing) and the E1 (missing fine steps and high stability). The twin-PBT (R75) and phasing-PBT (E1) combinations are analog methods of creating multiple filter bandwidths. *The R75 requires AM AGC modification.*

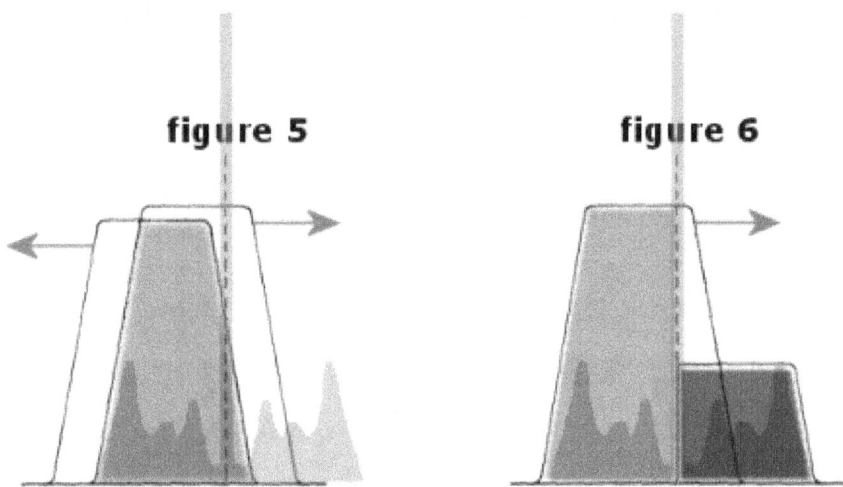

Figure 5: Twin-PBT (PBT movement depicted with arrows) selects LSB (left area is passed). Audio lows near the carrier (and the carrier itself) are partially attenuated. This can lead to *tinny* sound. **Figure 6:** Phasing-PBT selects LSB. The phasing unit eliminates USB (the black area).

IF-DSP receivers also provide multiple filters and hold much future promise. However, their current 8-bit "mechanical" sound is not ideal for SWL. The ADC's first 8-bits typically go for audio; the rest are for dynamic range (DR). At ~6 dB per remaining bit, 20-bits are needed for good DR.

Pete Gianakopoulos is a proponent of usage of low-NF (noise figure) moderate-IP3 mixers to avoid using IP3-degrading RF-amplifiers. Usually RF-amps are used to decrease NF (mixers are noisy devices compared to amplifiers); thus increasing MDS (sensitivity). Receivers with high-IP3 high-NF mixers (ex. RA6790GM's Quad-JFET) *require* RF-amps for optimal sensitivity. The TAK-3H (*Mini-Circuits*) is a level 17 mixer (+14 dBm @1 dB) with a conversion loss (approximates NF in a passive diode ring mixer) of *only* 4.82 dB. With no RF-amp the RA6790GM (6 kHz; 10 dB S/N) has a 2.5 µV sensitivity; a radio with an 8.5 dB composite NF (using a TAK-3H) is capable of ~0.3 µV.

7. OTHER TRICKS

Features, other than SAM-mode, can also be worked around when tuning DSBc stations in AM-mode. Amplified tuned-loops can lower NF, reduce unwanted mixer energy, and be rotated to attenuate local noise sources. A loop is a decent substitute for a random wire in urban locations.

Feature	Alternative
ss-SAM	SIDEBAND-SELECTED AM or PRECISION-ECSS.
PBT	Variable offset-tuning.
Notch	SIDEBAND-SELECTED AM; selecting the clean sideband.
Noise Blanker	Determine and eliminate the noise source.
DSP NR	Amplified tuned-loop antenna.
Tone Control	Variable offset-tuning with narrow filter.
IF DSP Filters	Twin-PBT or Phasing-PBT.

8. DISCUSSION

The SW community is reluctant to speak critically of SAM detection. I have owned several SAM units but most of my listening was done in AM-mode with an offset-tuned narrow filter and a slow-AGC (**SIDEBAND-SELECTED AM**). With harsh fading I used **PRECISION ECSS**. SSB-mode on the E1 and Sat800 sounded better than SAM-mode during carrier drops. Audio phasing insured that ECSS did not have the *tinny* sound sometimes associated with filter sideband selection. Since audio is disturbed less during carrier drops, ECSS is often superior to SAM during heavy selective fading.

Without knowing tricks one might assume 1-Hz tuning steps are overkill or that AM-mode is defenseless against selective fading distortion. While listening in AM-mode it is easy to assume that SAM-mode would have recovered audio which in truth was forever lost in a sideband fade.

It is hard to create an accurate, stable, and clean local oscillator. Of the current receivers only the $4250 RX-340 and $550 R75 have 1-Hz steps and ±1 ppm stability (for optimum ECSS). Ironically *Dave Zantow* calls the RX-340 SAM "*almost worthless*". *Dallas Lankford* calls the $1500 7030+ SAM "*unacceptable*". It is clear that even high-priced SAM units can whistle, hiss, distort, and have trouble gaining and holding lock. The $500 E1 contains a quality, Drake-designed SAM.

Considering the complexity needed to create SAM units, phasing units, and precision local oscillators, companies may wish to promote usage of **SIDEBAND-SELECTED AM**. I suggest adding an "**U**PPER, **C**ENTERED, **L**OWER" labeled button that shifts the local oscillator while keeping the displayed frequency constant. Usage would automatically select AM-mode, a narrow filter, and a slow-AGC.

The myth is that "SAM" alone reduces selective fading distortion. In reality it is a slow-AGC and bandwidth limiting (often suppression of one entire sideband) that is doing *much* of the work at reducing carrier dropout related distortion. Bandwidth limiting is possible through audio phasing (superior fidelity), IF filtering (**SIDEBAND-SELECTED AM** or **PRECISION ECSS**), and audio filtering (such as ELPAF). I wish to thank *Pete Gianakopoulos* for previewing this article.

2.9 English SW Broadcasts
The most active meter bands.
©2008

The chart below reflects the distribution of English SW broadcasts by meter band. Starting and ending band frequencies are the extended ranges given by *Passport*; frequencies consistent with FCC allocations are shown shaded. Using FCC ranges as much as 47% of the data was OOB (out-of-band). The **Use** column reflects when each band is active. Data is shown as percentages. **LIST** consists of 5542 pieces of data from six internet lists (n=204, 417, 543, 1343, 1503, 1532). Three columns use hourly data: 1) **TIME** sorted by time; **FREQ** sorted by frequency; and **NIGHT** representing 22:00 to 7:00 UTC (5 pm to 2 am EST). Nearly 150,000 data points were processed using several compiler BASIC programs; the chart was created using an MS Works spreadsheet.

Band	Start	End	Use	LIST	TIME	FREQ	NIGHT	Band
120M	2300	2495	–	0.16	0.00	0.00	0.00	120M
90M	3200	3400	–	1.32	1.54	1.11	1.27	90M
75M	3900	4050	–	0.54	0.00	0.00	0.00	75M
60M	4750	5100	Night	2.31	4.61	3.04	3.82	60M
* 49M	5730	6300	Night	16.94	44.17	28.67	36.56	49M
* 41M	6890	7600	Night	12.87	15.03	12.15	15.83	41M
* 31M	9250	10005	Night	21.20	20.47	20.37	16.94	31M
* 25M	11500	12200	Day	13.98	3.30	9.96	6.12	25M
* 22M	13570	13870	Day	6.48	1.28	8.28	4.45	22M
* 19M	15005	15825	Day	11.67	3.54	7.63	2.93	19M
16M	17480	17900	Day	6.98	0.85	3.79	0.71	16M
15M	18900	19020	Day	0.51	0.17	1.29	0.14	15M
13M	21450	21850	–	1.21	0.00	0.30	0.00	13M
11M	25670	26100	–	0.00	0.00	0.00	0.00	11M
OOB	–	–	–	3.83	5.04	3.41	11.23	OOB
n=	–	–	–	5542	35139	59254	42459	n=

Chart interpretation revealed that the vast majority of English SW broadcasts occur on the night bands of 49M, 41M, and 31M bands. These three bands, spanning from 5,730 to 10,005 kHz accounted for as much as 80% of English broadcasts. The bands 49M through 19M account for up to 88% of all English broadcasts. The **FREQ** column with nearly 60,000 individual hours of data is the most accurate. Here 49M to 31M and 49M to 19M account for 61% and 87% respectively. The analysis shows the importance of getting a radio with coverage outside the FCC designated bands. Anyone wanting to hear English SW broadcasts at night should use the 49M, 41M, and 31M bands. The data suggests that a homebrew SW receiver should minimally cover from ~5.7 to ~10.0 MHz.

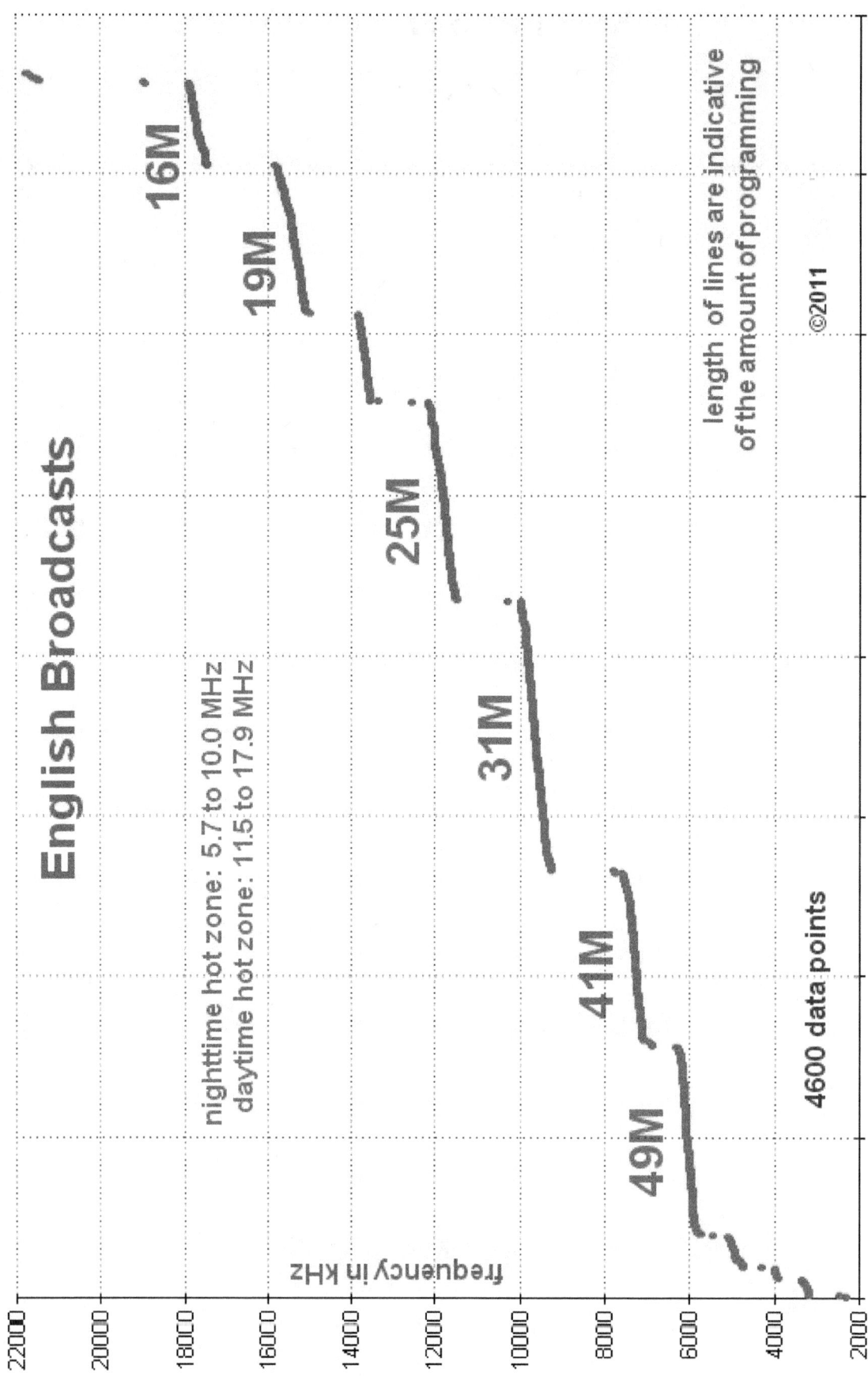

English Broadcasts

nighttime hot zone: 5.7 to 10.0 MHz
daytime hot zone: 11.5 to 17.9 MHz

16M

19M

25M

31M

41M

49M

length of lines are indicative
of the amount of programming

©2011

4600 data points

frequency in kHz

22000 20000 18000 16000 14000 12000 10000 8000 6000 4000 2000

3.1 Hacking the Si4734
Programming Handbook for Shortwave
©2010

Silicon Labs

This handbook distills 250 pages of Si4734 commands and properties down to 3 pages. It covers the fundamentals (setup; frequency and bandwidth changes; seeking) and various pitfalls.

Si4734 Setup			
Command 0x01	**POWER_UP**	byte2 = 0x11	byte3 = 0x05
Command 0x80	**GPIO_CTL**	byte2 = 0x0E	
Command 0x81	**GPIO_SET**	byte2 = 0x00	
Command 0x12	**SET_PROPERTY**		
byte2 = 0x00			
byte3 = Property HI		byte4 = Property LO	
byte5 = Value HI		byte6 = Value LO	
Property 0x3302	**AM_SOFT_MUTE_MAX_ATTENUATION**		
byte5 = 0x00	byte6 = 0x00		
Command 0x11	**POWER_DOWN**		

Command **0x01** powers the chip in AM mode with analog audio output, crystal oscillator via GPO3, non-patch boot, no CTS (Clear To Send) or GPO2 (General Purpose Output) interrupt. CTS typically takes ~300 µS while commands 0x01 and 0x12 take 110 and 10 mS respectively.

Command **0x80** takes the three output pins (GPO1, GPO2, GPO3) out of their hi-Z state. Command **0x81** sets the pins to ground. These two commands are necessary to reduce current and enhance stability. Unfortunately, chips prior to version 2.0 cannot execute these commands.

Command **0x12** allows us to set properties by sending six bytes containing the property and its new value. Setting property **0x3302** to zero disables soft muting; this facilitates DXing. Therefore, the six byte sequence: 0x12, 0x00, 0x33, 0x02, 0x00, 0x00 will disable soft muting.

Command **0x11** causes the chip to enter a low-power mode that only responds to **0x01**. In this mode the chip consumes 10 µA. Typical AM current is 17.3 mA: using 52 mW at 3 Volts. Switching between AM and FM modes requires **POWER_DOWN** (all register contents are lost). Version 1.0 requires a reset if sent a command other than POWER_UP while in POWER_DOWN.

An MCU needs only three lines (SCLK, SDIO, RST or clock, data, reset) for control of the Si4734. There is one big **caveat**: all data on the SDIO line (meaning MCU control of the Si4734, or any other device on the bus) must be suspended during tuning and seeking. Failure to do so may result in crystal oscillator jitter and lead to mistuning, false seek stops, and a lowered SNR. Radios should not attempt to poll signal strength information (SNR/RSSI) during tuning/seeking.

Due to the above warning, advanced control programs must setup GPO2/INT (PIN 18) as an interrupt line. This is accomplished by setting byte2 of POWER_UP to 0xD1 [bit 6 for INT and bit7 for CTS]. During interrupts INT is driven low for 1 µS. The STCINT or Seek / Tune Complete Interrupt should be used instead of polling. **GPO_IEN** configures interrupt resources, including: STCINT, CTS, ERR (Error), ASQINT (Audio), and RSQINT (Received Signal Quality). To examine and update the interrupt status byte, **GET_INT_STATUS** is called. **AM_RSQ_INT_SOURCE** is used to create interrupts based on signal quality: SNR or RSSI (triggered by low or high values).

Si4734 Frequency and Bandwidth
Command 0x40　　**AM_TUNE_FREQ**
byte2 = 0x00
byte3 = Frequency HI　　byte4 = Frequency LO
byte5 = Capacitor HI　　byte6 = Capacitor LO
Property 0x3102　　**AM_CHANNEL_FILTER**
byte5 = 0x00　　byte6 = 0x00 - 0x04

Command **0x40** permits tuning both the I/Q mixer LOs and the onboard capacitor. Valid capacitor values are 0 (automatic tuning) and 1 to 6143. Capacitor values can be calculated via the formula: **0.095 * word_value + 7** in pF. This calculates to a tuning range of ~7 to 590 pF.

Property **0x3102** allows setting filter bandwidth. Values include: 0x00 6-kHz, 0x01 4-kHz, 0x02 3-kHz, 0x03 2-kHz, and 0x04 1-kHz. Note: the 1-kHz filter requires version 2.A.2 or newer.

Si4734 Seeking
Property 0x3400　　**AM_SEEK_BAND_BOTTOM**
byte5 = Frequency HI　　byte6 = Frequency LO
Property 0x3401　　**AM_SEEK_BAND_TOP**
byte5 = Frequency HI　　byte6 = Frequency LO
Property 0x3402　　**AM_SEEK_FREQ_SPACING**
byte5 = 0x00　　byte6 = 0x05
(0x01,0x09,0x0A)
Property 0x3403　　**AM_SEEK_TUNE_SNR_THRESHOLD**
byte5 = 0x00　　byte6 = 0x05 — 0x3F
Property 0x3404　　**AM_SEEK_BAND_RSSI_THRESHOLD**
byte5 = 0x00　　byte6 = 0x05 — 0x3F
Command 0x41　　**AM_SEEK_START**
byte2 = 0x00 + 0x08 [UPWARD] + 0x04 [WRAP]
Command 0x42　　**AM_TUNE_STATUS**
byte2 = 0x02

Properties **0x3400** and **0x3401** setup the seek range in kHz (149 to 23000 being valid). While **0x3402** sets the spacing to 5-kHz (other applicable values are 1-kHz, 9-kHz, and 10-kHz).

Properties **0x3403** and **0x3404** set the SNR (Signal to Noise Ratio) and RSSI (Received Signal Strength Indicator) values in dB and dBμV respectively. Range is 1 to 63; 0 being disabled.

S-meter Equivalents		
S1	0.2 μV	−14 dBμV
S2	0.4 μV	−8 dBμV
S3	0.8 μV	−2 dBμV
S4	1.6 μV	4 dBμV
S5	3.2 μV	10 dBμV
S6	6.3 μV	16 dBμV
S7	13.0 μV	22 dBμV
S8	25.0 μV	28 dBμV
S9	50.0 μV	34 dBμV
S9+10	160.0 μV	44 dBμV
S9+20	500.0 μV	54 dBμV

Command **0x41** starts seeking downward and halts at the end of the band. Adding 0x08 causes upward seeking and adding 0x04 causes wrapping around when at the end of the band.

Command **0x42** is used to determine <u>where</u> the radio stopped. Value 0x02 aborts seeking. It takes 80 mS per station to seek. Seeking the 31M band (9250 to 9995 kHz) takes ~59 seconds.

Response Bytes	
GET_PROPERTY	3
AM_RSQ_STATUS	5
AM_TUNE_STATUS	7
GET_REV	8
POWER_UP	15

Note: POWER_UP returns 15 bytes only in Query Library ID mode.

All of the commands in the "Si4734" tables above return <u>zero</u> response bytes: except for the **AM_TUNE_STATUS** command; which is used to abort seeking. This fact can be used on simple control programs along with the general timing guidelines. The chip uses an I2C bus as default: this can reach speeds of 400 kHz. The 3-wire and SPI buses can run as fast as 2.50 MHz. Using SEN (PIN 6) [Serial Enable] the I2C bus address can be altered: 0x22 [SEN LO]; 0xC6 [SEN HI].

The following may help simplify some programs. It is possible to send all seven argument bytes for all commands. However, unused arguments must be sent as 0x00. It is also possible to read all fifteen response bytes for all commands. However, all unused response bytes will contain random values. WARNING: writing a block of more than eight bytes (one command byte followed by 7 argument bytes) to the chip will result in an input buffer overrun and unpredictable behavior.

Other Si4734 Commands		
Command 0x10	**GET_REV**	Read Revision Data
Command 0x13	**GET_PROPERTY**	Read Property Values
Command 0x14	**GET_INT_STATUS**	Read Interrupt Status
Command 0x15	**PATCH_ARGS**	Software Patch (15856 bytes maximum)
Command 0x16	**PATCH_DATA**	Software Patch (~0.5 second via I2C)
Command 0x43	**AM_RSQ_STATUS**	Received Signal Quality

Interrupt related function calls are highlighted.

Other Si4734 Properties		
0x0001	**GPO_IEN**	Interrupt Enabling
0x0201	**REFCLK_FREQ**	31130—32768 Hz SW (0 disables AFC)
0x0202	**REFCLK_PRESCALE**	1-4095 divisor (support to 40 MHz)
0x3100	**AM_DEEMPHASIS**	50 μs
0x3200	**AM_RSQ_INT_SOURCE**	Interrupt Source:
0x3201	**AM_RSQ_SNR_HIGH_THRESHOLD**	SNR > 0—127 dB
0x3202	**AM_RSQ_SNR_LOW_THRESHOLD**	SNR < 0—127 dB
0x3203	**AM_RSQ_RSSI_HIGH_THRESHOLD**	RSSI > 0—127 dBμV
0x3204	**AM_RSQ_RSSI_LOW_THRESHOLD**	RSSI < 0—127 dBμV
0x3300	**AM_SOFT_MUTE_RATE**	1-255 * 4.35 dB/s
0x3301	**AM_SOFT_MUTE_SLOPE**	1-5 dB/dB
0x3303	**AM_SOFT_MUTE_SNR_THRESHOLD**	0—63 dB
0x4000	**RX_VOLUME**	0-63 (default 63)
0x4001	**RX_HARD_MUTE**	Mute Audio Output

Interrupt related variables are highlighted.

Silicon Labs

Si4734 Frequency Coverage			
LW	153– 279 kHz	talk, Europe, Africa	2800 µH
MW	520–1710 kHz	talk,	180–450 µH[1]
SW	2.3– 23 MHz	global, China, India, Brazil	
FM	64– 76 MHz	college, emergency, TV	
FM	76– 108 MHz	global	

[1]*Mating a 15 µH air loop to a 1:5 ratio inductor yields 375 µH.*

Silicon Labs is revolutionizing radio with their Si4734: a CMOS AM/FM/SW/LW low-IF DSP receiver chip. This IC is part of a new era wherein portables will no longer need ceramic IF filters. Radio design will instead focus on the antenna circuit, audio amplification (ex. Sony CXA1622M), user interface, and MCU control. These radios will require little or no manual (factory) alignment.

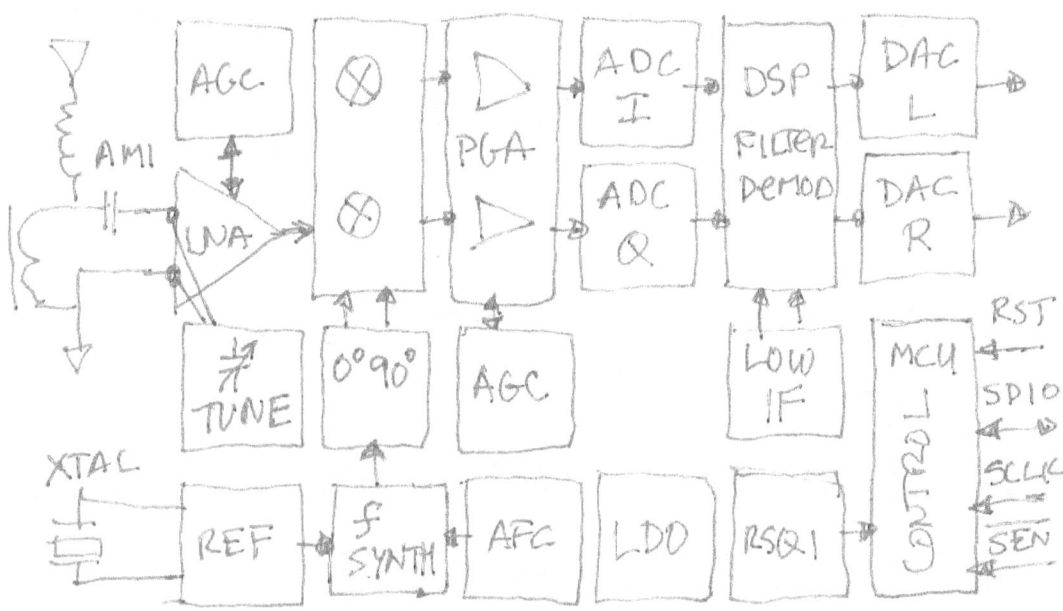

Tuned RF energy (see below) enters a low noise amplifier (LNA) whose gain is under MCU control. The automatic gain control (AGC) can also adjust the programmable gain amplifier (PGA) that follows each mixer. The I/Q (Inphase/Quadrature) lines are digitized using low- or band-pass high-resolution analog to digital coverts (ADC). A digital low-IF (intermediate frequency) signal is presented to a digital signal processor (DSP) for bandwidth limiting and demodulation. This digital section is 24-bit and capable of sampling at 48-kHz. A digital to analog converter (DAC) creates a high impedance (10k-ohm) audio signal (THD 0.1%) to be boosted by an external audio amplifier. The Si4734 also contains a low dropout voltage regulator (LDO) [two 1.5 Volt batteries can supply both VDD and VIO], an automatic frequency control (AFC), a radio signal quality interrupt (RSQI) line, and a separate FM section [includes FM input line (PIN 2; FMI), FM LNA, and FM I/Q mixers].

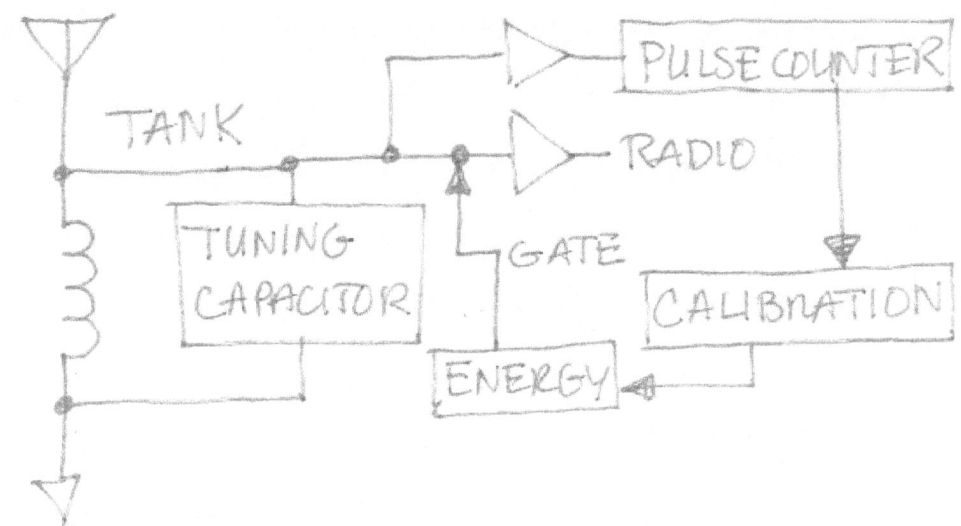

One unique feature of the Si4734 is its ability to <u>automatically</u> tune AM antennas. During AM operation the antenna's external inductor forms a tank with the chip's internal capacitor. The (calculated) tuning range of the internal capacitance is from 7.095 to 590.585 pF. The low noise amplifier input is of high impedance. The capacitor value can also be <u>manually</u> set during tuning. Multiple weighted parallel MOS capacitors, each with a transistor switch, are under 13-bit digital control. The chip enters a calibration mode when a new station is tuned. Under MCU control (see above), energy is added to the tank by forcing current in its inductor (DC is converted to AC). This energy dissipates via damped oscillations (ringing) at the tank's resonant frequency. A pulse counter (fed by a high input impedance amplifier) determines the frequency of the ringing. The MCU iteratively adjusts the IC's tunable capacitance until the resonant frequency of the tank equals the desired (seeked or tuned) frequency. I wish to thank *Scott Willingham* of Silicon Labs.

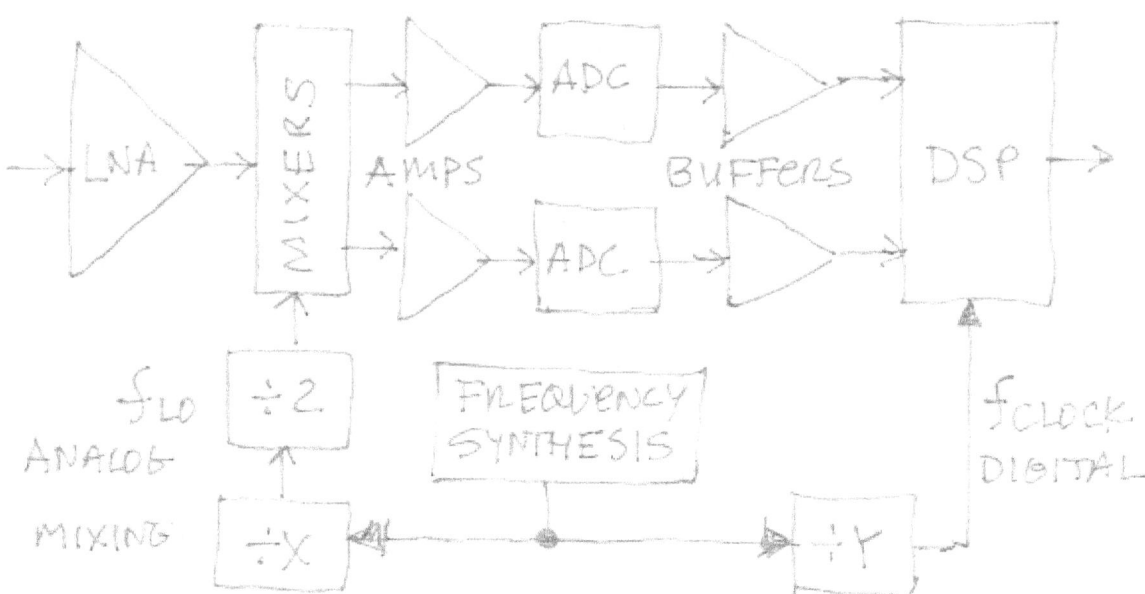

The Si4734 also contains a ratiometric clock, meaning a single oscillator is divided down for use by both the local oscillator (divide by 2 providing I/Q signals) and the digital circuitry. In doing this, digital harmonics fall on the frequency of the oscillation signal. This feature permitted Silicon Labs to pack a DSP, ADC, DAC, MCU, and other clocked digital components on a chip that contains LO generation and RF mixers. The chip is capable of low and high-side LO injection. The design supports a programmable IF. The chip is capable of correcting for frequency inaccuracies. The Si4734 is firmware upgradeable and up to **15,856** bytes can be written as a software patch.

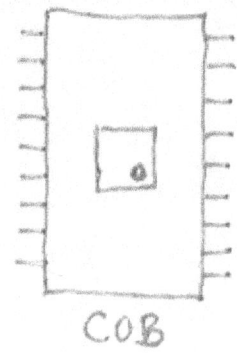

COB

The Si4734 contains advanced wireless communications technology. It has incorporated ESD (electrostatic discharge) protection (two input protection diodes) for its low noise amplifier (LNA). Its LNA has a digitally controlled attenuation network (variable bias resistance). The chip contains a digital automatic gain control (AGC). Peak detection circuits are located after the LNA and each VGA. A digital processor (apart from the main DSP) uses the detectors (set to high and low thresholds) to incrementally adjust the gain. If the RSSI (Received Signal Strength Indicator) does not track the change in gain of the LNA (1 dB RSSI per dB gain) then a flag is set indicating possible intermodulation distortion. The MCU can correct the situation by reducing amplifier gain. The gain is adjusted from the antenna (LNA) inward. The ADC sets an overload line (O/L) when a repetition of output occurs over several clock cycles (gain being high, SNR low, noise floor high). The Si4734 has an unconditionally stable ADC based on a delta-sigma modulator (DSM). Its ADC can maintain stability regardless of the amplitude or frequency of the incoming signal. DSM was used to permit high resolution conversion by means of a low-cost CMOS manufacturing process.

34 = Si4734   1B = Firmware
B = Die Rev   3NS = Tracking
● = Pin 1   7 = 2007 Year
                  39 = 39th Week

The 20-pin Si4734 is 3 mm square and housed in a QFN package. Manufacturers also use chips that are mounted on an 18-pin board or COB (chip on board) assembly. Each chip has YAG laser surface markings. The "34" denotes a Si4734 and is followed by the firmware version. Next is the die revision, here "B", followed by a three digit tracking code. The circle denotes PIN 1 and is followed by year "7" (meaning 2007) and by 2 week digit codes "39" (meaning the 39th week).

The sensitivity specification for the Si4734 on AM is given as: 25 μV for 26 dB (S+N)/N. It was calculated that the receiver chip, at 3 dB (S+N)/N, has a sensitivity of 1.77 μV or about **S4**. An **S2** (0.4 μV) signal can be heard with +12.9 dB gain: this is achievable via toroid turns ratios.

3.3 Hacking the Si4734
Hellenized Si4734 Shortwave Receiver
©2010

Silicon Labs

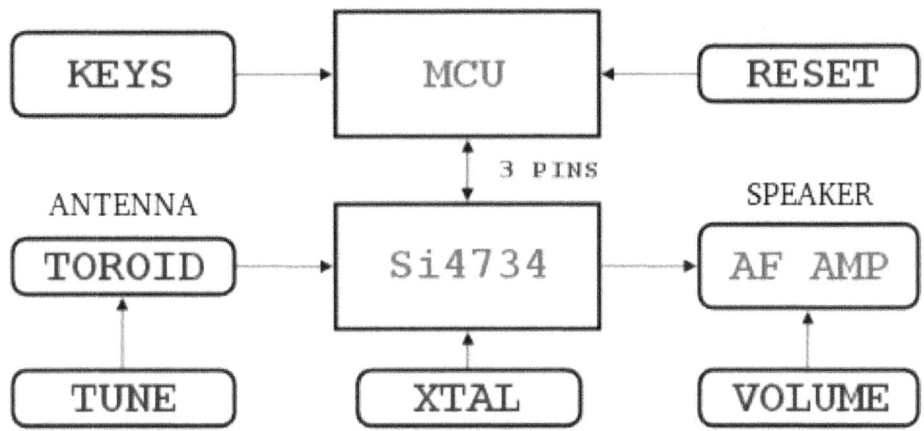

Above is a block diagram for a Si4734-based shortwave receiver. The goal was a no-frills shortwave DX machine. There is no display: punch in a frequency, select a filter, set the volume, peak the antenna, and listen. Hardware was broken into three sections: MCU, Si4734, and audio amplification. Software sections include: keypad polling, boot-up, plus bandwidth and frequency changing. **WARNING**: this is a preliminary design and may contain hardware or software errors.

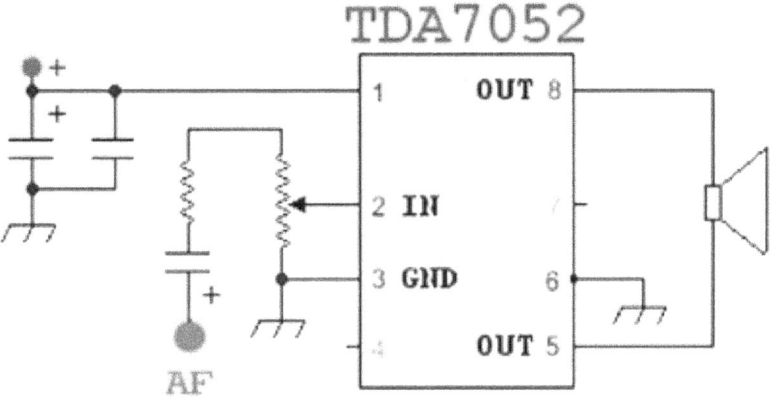

The audio amplification section consists of a TDA7052 1 Watt BTL (Bridge Tied Load) Mono Audio Amplifier. Pin 1 is power, decoupled via two capacitors. Pin 6 is ground. Pins 5 and 8 drive an **8-ohm** speaker. Pin 2 accepts audio from the Si4734 through a potentiometer volume control. Pin 3 is the input's ground. The TDA7052 is available as a PCB kit and is shown, assembled, below.

61

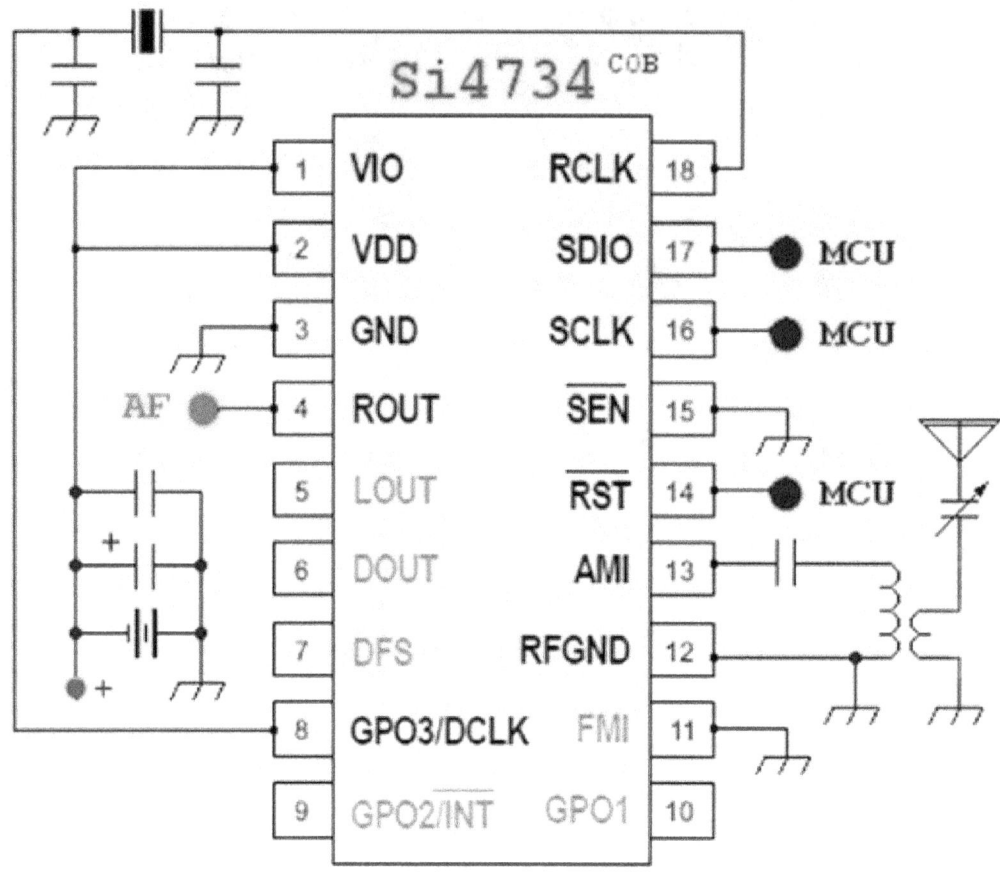

The heart of the radio is Silicon Lab's Si4734 low-IF DSP receiver IC [on a COB assembly]. Power is derived via two "D" type batteries, decoupled by two capacitors. This is applied to Pins 1 (digital I/O voltage) and 2 (supply voltage). Pin 3 is ground. Pin 4, the right audio output, is sent to the TDA7052 chip for amplification. Pins 5, 6, and 7 (left output, digital lines) are unconnected. Pins 8 and 18 are connected via a 32.768 kHz crystal. Pins 9 and 10 (outputs) are not connected. Pin 11, the FM input, is grounded. Pins 12 and 13 form the AM input and are coupled via a toroid to an antenna with a trimmer capacitor. Pins 14, 16, and 17 are used for MCU control: reset, I2C clock, and I2C data lines; respectively. Pin 15 is grounded, setting the I2C **bus address** to 0x22.

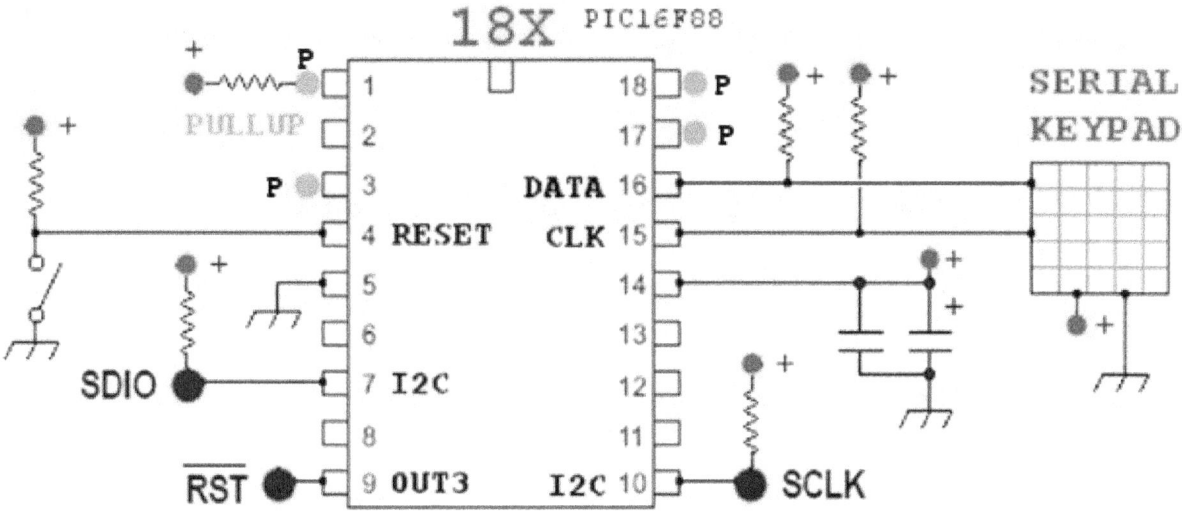

The MCU is a PICAXE 18X. Pin 14 is power, decoupled via two capacitors. Pin 5 is ground. Pin 4 is a reset. Pins 7, 9, and 10 control the Si4734 IC via its I2C data, reset, and I2C clock lines. Pins 15 and 16 receive serial keypad data. Pins 1, 3, 17, and 18 (and others) use pull-up resistors. A substitute to a serial keypad, a **4 by 3** keypad, can be polled using a single **ADC** input (Pin 17).

The user interface is a keypad. Three commands are possible: 9 ENTER reboots, 0 thru 8 ENTER changes bandwidth, and typing a 3 to 5 digit number then ENTER changes the frequency.

Software consists of a loop that polls keypad input and assembles a number. When ENTER is pushed, the program determines if the user requested a reboot, filter change, or tuning change.

```
                       MINIMALIST Si4734 SOFTWARE
main:          ; initialize variables
w0 = 0 | w1 = 0 | w2 = 100000              ; w2 is 10's place
polling:       ; keypad polling
KEYIN                                      ; get a key
w1 = -5                                    ; w1 holds digit
IF keyvalue = $45 THEN w1 = 0             ; check for 0 to 9
IF keyvalue = $16 THEN w1 = 1
IF keyvalue = $1E THEN w1 = 2
IF keyvalue = $26 THEN w1 = 3
IF keyvalue = $25 THEN w1 = 4
IF keyvalue = $2E THEN w1 = 5
IF keyvalue = $36 THEN w1 = 6
IF keyvalue = $3D THEN w1 = 7
IF keyvalue = $3E THEN w1 = 8
IF keyvalue = $46 THEN w1 = 9
IF keyvalue = $5A THEN enter              ; check for ENTER
IF w1 = -5 THEN polling                   ; is key invalid?
w2 = w2 / 10                              ; next 10's place
w0 = w1 * w2 + w0                         ; add digit to w0
GOTO polling
enter:         ; interpret command
IF w2 = 0 THEN main                       ; divide by zero?
w0 = w0 / w2                              ; fix 10's place
IF w0  =   9              THEN boot       ; boot
IF w0 >=   0 and w0 < 9   THEN filter     ; filter
IF w0 >= 149 and w0 < 23001 THEN tune     ; tune
GOTO main

boot:          ; initialize Si4734
LOW 3 | PAUSE 250 | HIGH 3 | PAUSE 250    ; hard reset
I2CSLAVE $22,i2cfast,i2cbyte             ; setup I2C
WRITEI2C $22,($01,$11,$05)               ; POWER_UP
WRITEI2C $22,($80,$0E)                   ; GPIO_CTL
WRITEI2C $22,($81,$00)                   ; GPIO_SET
WRITEI2C $22,($12,$00,$33,$02,$00,$00)   ; SOFT_MUTE_OFF
GOTO main

filter:        ; bandwidth change
WRITEI2C $22,($12,$00,$31,$02,$00, b0)   ; AM_CH_FILTER
GOTO main

tune:          ; frequency change
WRITEI2C $22,($40,$00, b1, b0,$00,$00)   ; AM_TUNE_FREQ
GOTO main
```

3.4 Hacking the Si4734
Shortwave Antenna Modification
©2010

Silicon Labs

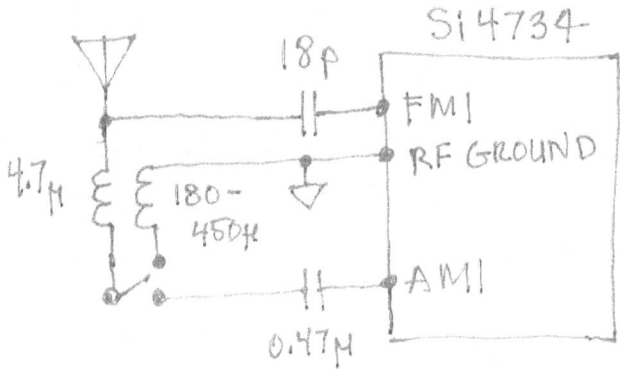

The critical ingredient in any Si4734-based design is the antenna system. The FM IP3 was given as 105 dBμV at 2-MHz spacing. This calculates to only -2.0 dBm at 50-ohms. Unwanted RF energy must be kept out of the LNA to prevent mixer overload. Manufacturers tend to follow the available datasheet designs. Silicon Labs' schematics showcase the need for a minimum amount of external components. However, it is better to use a tuned circuit than a wideband collector of shortwave energy. Note: the 0.47 μF capacitor is for DC isolation (large, as not to affect tuning).

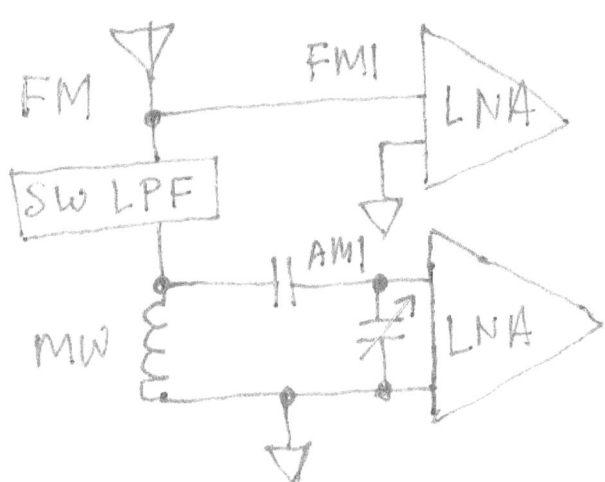

Above is the block diagram of the Tecsun PL-310, as per their manual. The antenna is not ideal for shortwave: RF energy from the whip is low-pass filtered (perhaps an inductor) and then sent to the MW tank. If the chip's capacitor were set to 590 pF then much of the SW energy would be lost: 3 MHz sees 590 pF as 90 ohms and 23 MHz sees it as 12 ohms. **Speculating**: the Tecsun engineers would be forced to set the capacitor to its minimum value of 7 pF (0x0001) to minimize losses. If MW ferrite inductance is 330 μH then the tank is resonant at 3.3 MHz. Replacing the rod with a small inductor (4 μH) may not aid SW performance: software setting of the capacitance to 0x0001 is untuned and disables automatic tuning. It may be inviable to try sw antenna variations without software hacking. Manufacturers may wish to incorporate both <u>an automatic and a manual</u> capacitance setting, as well as PIN diodes for antenna selection, and a Si4734 computer interface.

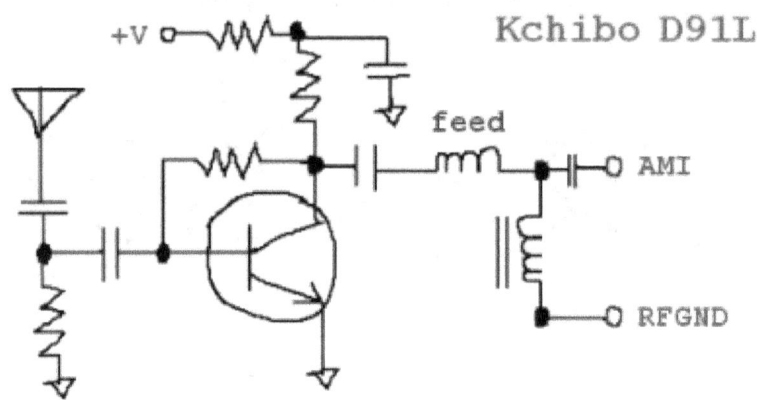

Kchibo D91L

Inspection of the PCB of the Kchibo D48L and D91L revealed common emitter transistor amplified shortwave antennas. SW energy is fed into the MW ferrite tank via an inductor (same topology as the PL-310). Unfortunately, the Kchibo design exposes the LNA/mixers to amplified broadband (MW/HF) RF energy; plus MW energy captured by the ferrite. Local BCB stations are likely to cause overload on the shortwave bands. Feeding amplified RF into the chip's LNA could push it into its non-linear region. The end result may be a sensitive radio with spurious signals.

Antenna Toroid Calculations			
T37-2	32-turns	4.096 μH	3.24-27.80 MHz
T50-2	29-turns	4.121 μH	3.23-27.72 MHz

Note: stray antenna capacitance will hinder tuning the upper range.

A red iron-powder toroidal inductor of type 2 material (1 to 30 MHz frequency range) can be used to form a tank with the chip's internal tuning capacitor. A toroid is compact, cheap, high "Q", and fairly self-shielding. The proposed shortwave antennas below **were not yet tested** but are based on regenerative receiver designs. Although shown with no ferrite rod, **PIN** diodes (ex. HP 5082-3080) or a rotary switch can be used to select between MW and SW. It may be feasible to add an external low-range-enhancing capacitor in parallel to the Si4734's internal capacitance.

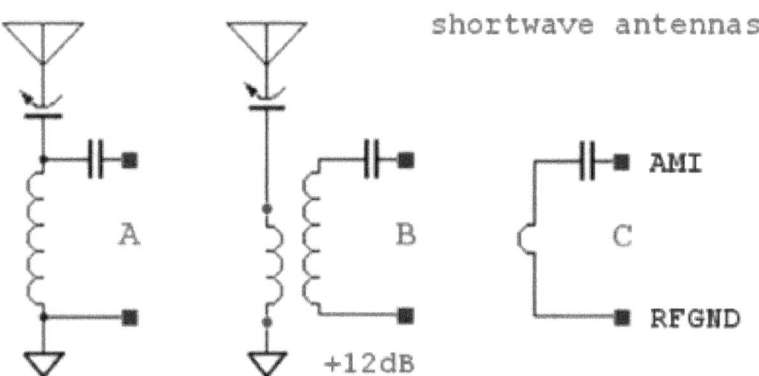

shortwave antennas

The inductors above form a tank with the chip's internal capacitor. Antenna **A** uses a trim capacitor to introduce shortwave energy to the tank; this prevents loading. Antenna **B** increases voltage via the turns ratio on the toroid. A trim capacitor negates the need for taps. Air-coupled, this setup may utilize a tuning cap (circular taps). Antenna **C** is a 30" by 3" single turn metal loop.

3.5 Hacking the Si4734
Kchibo KK-D48L Circuit Description
©2010

Silicon Labs

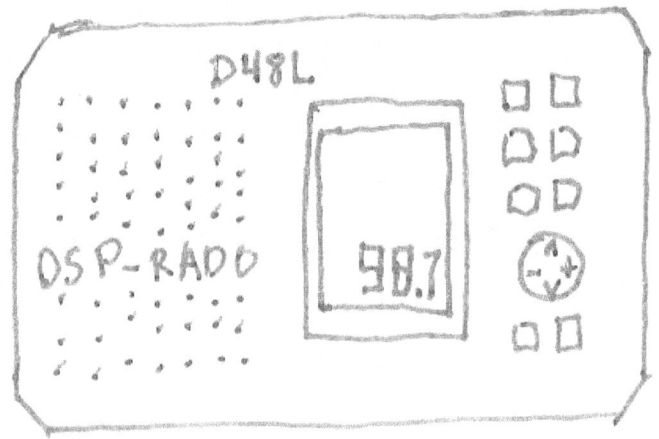

The Kchibo KK-D48L is a shortwave receiver based on Silicon Labs' Si4734 chip. This IC contains an automatic antenna tuner, I/Q mixers for image rejection, low-IF DSP filtration and demodulation, and a microcontroller. The D48L is 4.8 ounces and takes up ~11.9 cubic inches.

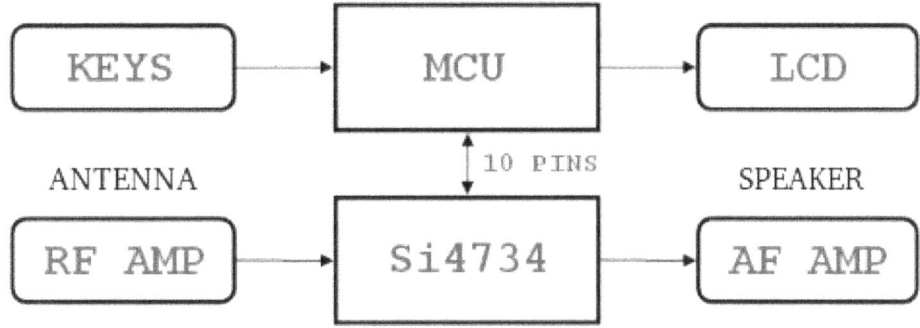

The MCU board consists of a microcontroller and user interface (keypad and display). The DSP board houses the Si4734 tuner chip, radio frequency (RF), audio frequency (AF), and power circuitry. The two boards are attached to one other via a 10-pin connector, allowing MCU control.

The MCU board consists of a ~48-pin MCU on a COB assembly (90 memories), tuning fork, three SMT transistors, and 220 µF decoupling capacitor. This section also houses inputs including: 12 buttons (4 housed under a circle), reset dimple, and a light sensor. In addition this half of the radio contains outputs: a backlit LCD display, power indicator LED, and 16-ohm 250 mW speaker.

12 BUTTON INPUTS
Power, Volume, Key Lock
Memory Store, Memory Address, Memory Scan
Band (FM/MW/SW), Meter Band (SW)
Up, Down, Plus, Minus

LCD DISPLAY OUTPUT
Frequency, MHz/kHz, MW/FM/FM-Stereo, SW Meter Band
SNR in dB, Signal Level in dBµV, Memory
Volume, Lock, Time, Date, Sleep
Power Level, External Power

The DSP section houses a 3.7 Volt lithium-ion battery with a 470 µF decoupling capacitor, and external power jack. The MCU section receives power via PIN_4, through a protection diode. It receives ground via PIN_9, which comes from the ground strip located underneath the Si4734. RF amplifier and Si4734 power is MCU dispersed (PIN_1) and controlled (PIN_2) via a transistor. The Si4734 receives power through an RF choke (preventing RF from entering) and is decoupled via a 1000 µF electrolytic capacitor. The MCU controls (PIN_3) power to the AF amplifier through transistor switching. Hence, the MCU can shut down the DSP board for low-power (sleep) mode.

The antenna section of the DSP board contains three inputs: a 3.6" MW ferrite rod, an FM and SW aerial, and an external antenna jack. The MW ferrite rod input is fed into the Si4734 AMI (via a DC isolation capacitor) and RFGND pins. The FM / SW inputs are diode protected and must pass a small capacitor (RF-only). The RF energy is then transistor amplified in a common-emitter configuration. The amplified signal undergoes LC **T-section** low pass filtering (excluding UHF and above) before entering the Si47347's FM input or FMI pin. The amplified signal can also be switch fed (shielding MW) via the Si4734's GPO1 pin (under direct MCU control) into the MW ferrite tank via an inductor. This inductor acts as a VHF RF choke and prevents VHF and higher from entering.

The Si4734 chip has its SEN (serial enable) pin grounded. The chip is under complete MCU control via four connector pins: PIN10 to RST (reset), PIN_6 to RCLK (main clock), PIN_8 to SCLK (data clock), and PIN_7 to SDIO (data input/output). The Si4734 uses an **I2C** bus: address 0x22.

The audio section consists of a SA1622 Stereo Power Amplifier chip. It receives input from the Si4734's LOUT and ROUT pins. The chip has a 1000 µF decoupling capacitor and 100 µF ripple capacitor. The SA1622 VOLUME pin is set to a <u>fixed</u> value (meaning volume is MCU controlled via the Si4734 IC). The SA1622 SWITCH pin has <u>two</u> inputs: it is in BTL (bridge tied load) mode with the internal speaker and stereo mode with an external headphone. However, the MCU can also put the chip into BTL mode via PIN_5 whenever the signal level drops below 25 dBµV, improving SNR.

10-PIN CONNECTOR	
PIN_1: Power Si4734/RF-Amplifier	PIN_6: Si4734 RCLK (main clock)
PIN_2: Power Si4734/RF-Amplifier	PIN_7: Si4734 SDIO (data I/O)
PIN_3: Power Audio Amplifier	PIN_8: Si4734 SCLK (data clock)
PIN_4: Power +3.7 Volt	PIN_9: Ground
PIN_5: Force FM Mono	PIN10: Si4734 RST (reset)

The D48L is an admirable design; however, there is room for improvement. Software wise, the radio should disable soft muting and include 1-kHz tuning steps. The SW antenna setup could be improved with a toroidal tank tuned by the Si4734's capacitor. The Si4734 should ideally be setup to utilize its interrupt line GPO2/INT; and be clocked via its own fork and GPO3/DCLK. Also, volume is controlled better using a potentiometer at the SA1622, not via the Si4734. The ground plane on the DSP board should be more extensive and there is no RF shielding on the MCU board.

3.6 Si4734 Digital Filters
6, 4, 3, 2, and 1 kHz
©2011

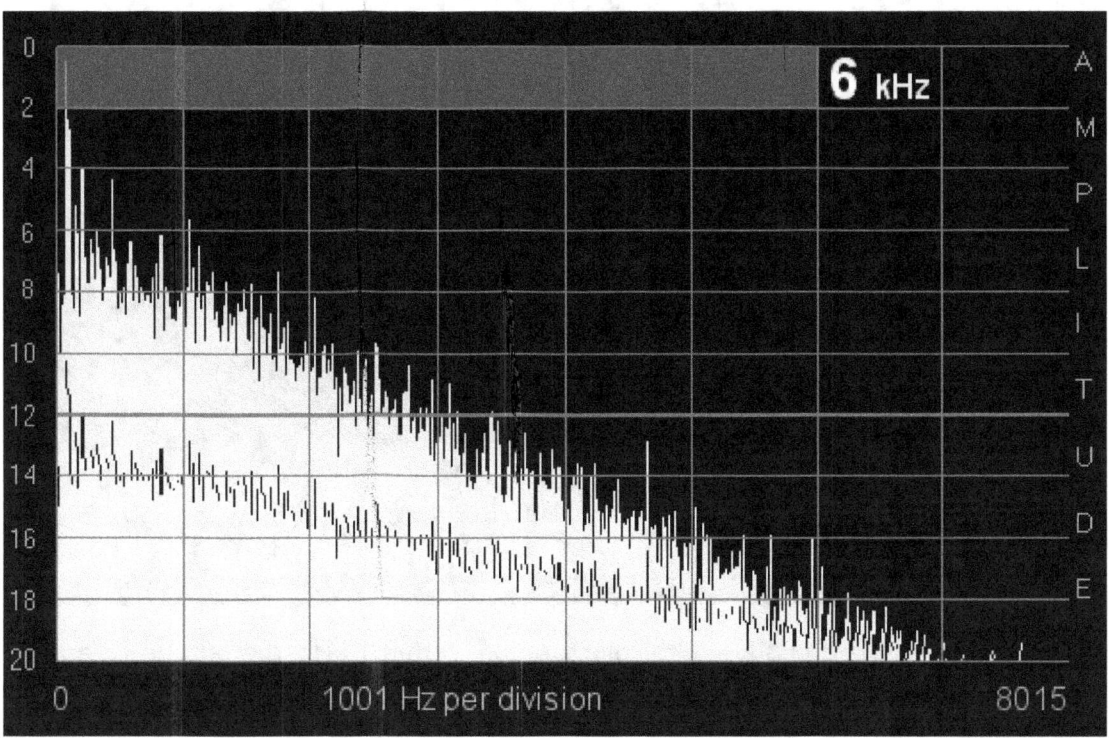

The graph above shows the attenuation of background noise via the 6-kHz filter of Silicon Lab's Si4734. The audio of a Tecsun PL-390 was fed into a PC sound card, at a 48,000 Hz sample rate. The horizontal axis is linear: 1001 Hz per division. The vertical axis has a log 10 scale: 2 dB per division. The gray traces represent peeks over a timed period. By 7 kHz there is no output.

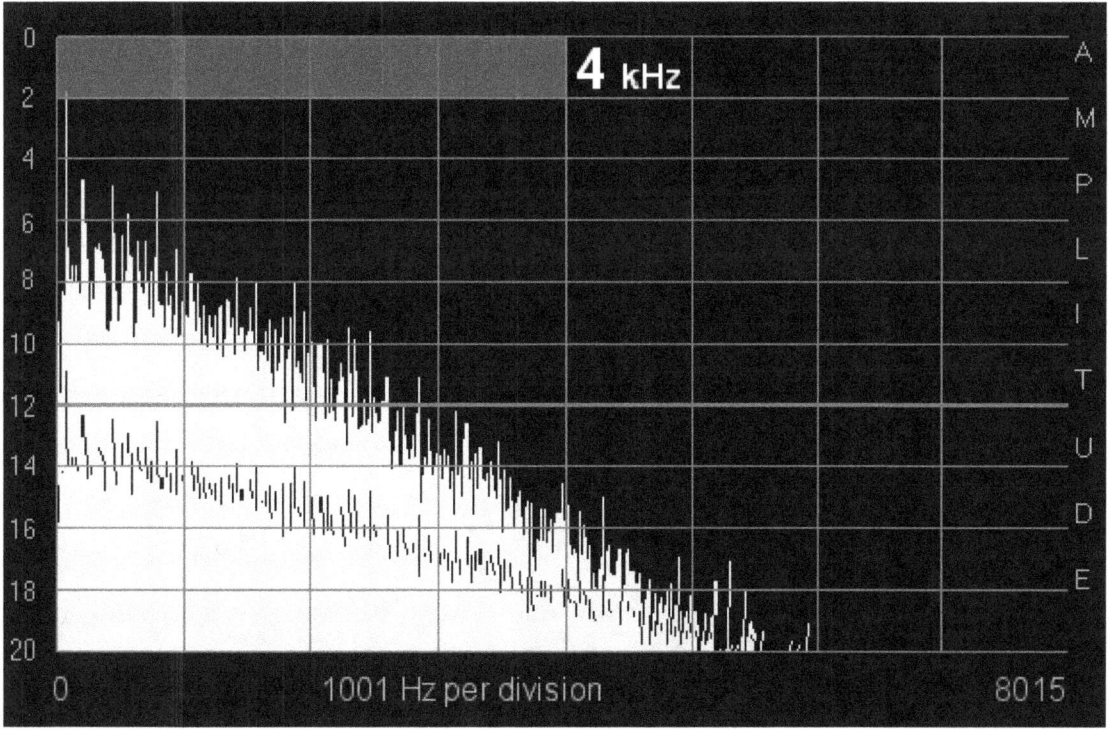

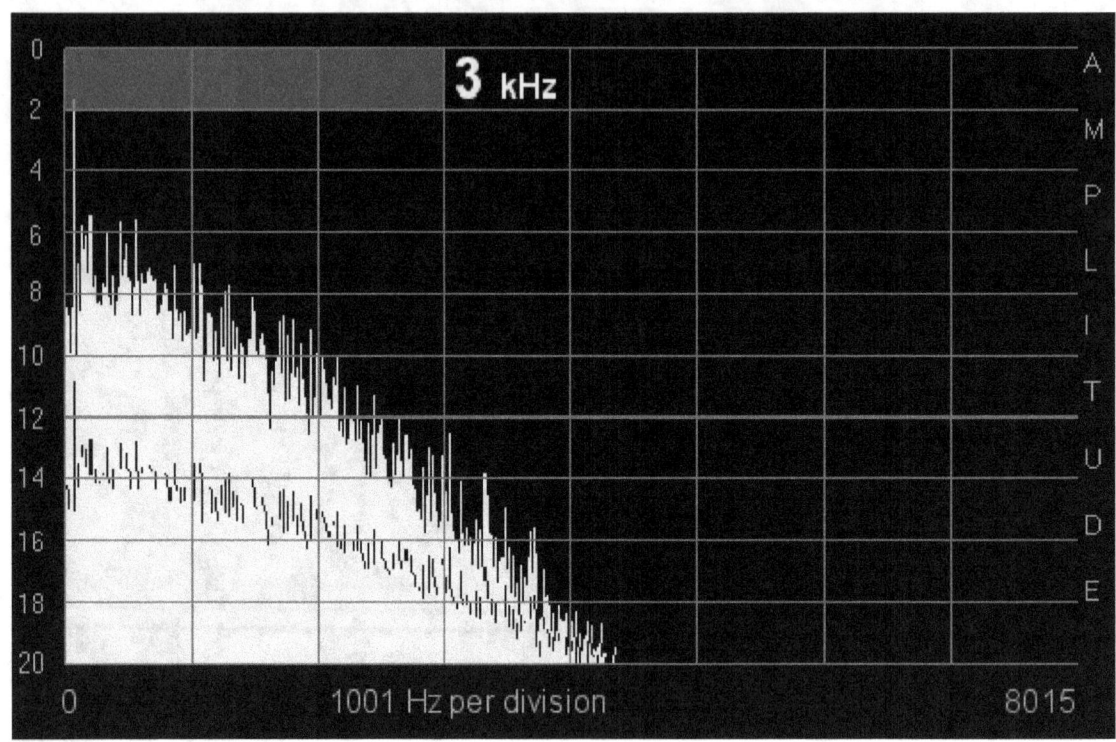

Above are the 4-kHz and 3-kHz filter plots. Again, within 1 kHz, the output is nearly zero. At these bandwidths this tight of a shape factor is beyond that of ceramic, mechanical, or crystal filters. Below are the 2-kHz and 1-kHz filter plots. Within ~1 kHz all output has been attenuated. The last graph shows all 5 Si4734 digital filter bandwidths superimposed over top of one another.

The *Sony CXA1129N*, in the SRF-59, allowed consumers a taste of low-IF in an analog MW radio. And now, the *Silicon Labs Si4734*, in the PL-390 and other radios, has given consumers a taste of low-IF DSP in a digital MW, FM, and SW radio. Silicon Labs, of Austin, Texas, is now the high technology leader in the shortwave radio industry. Their chips are a huge win for consumers.

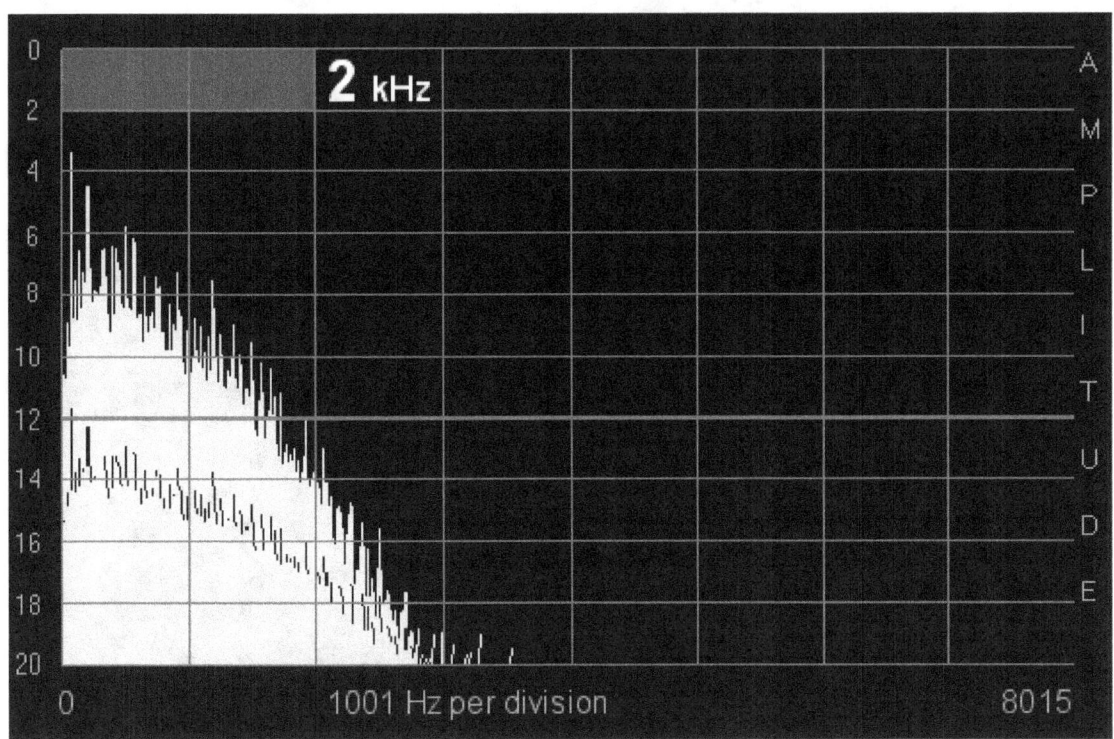

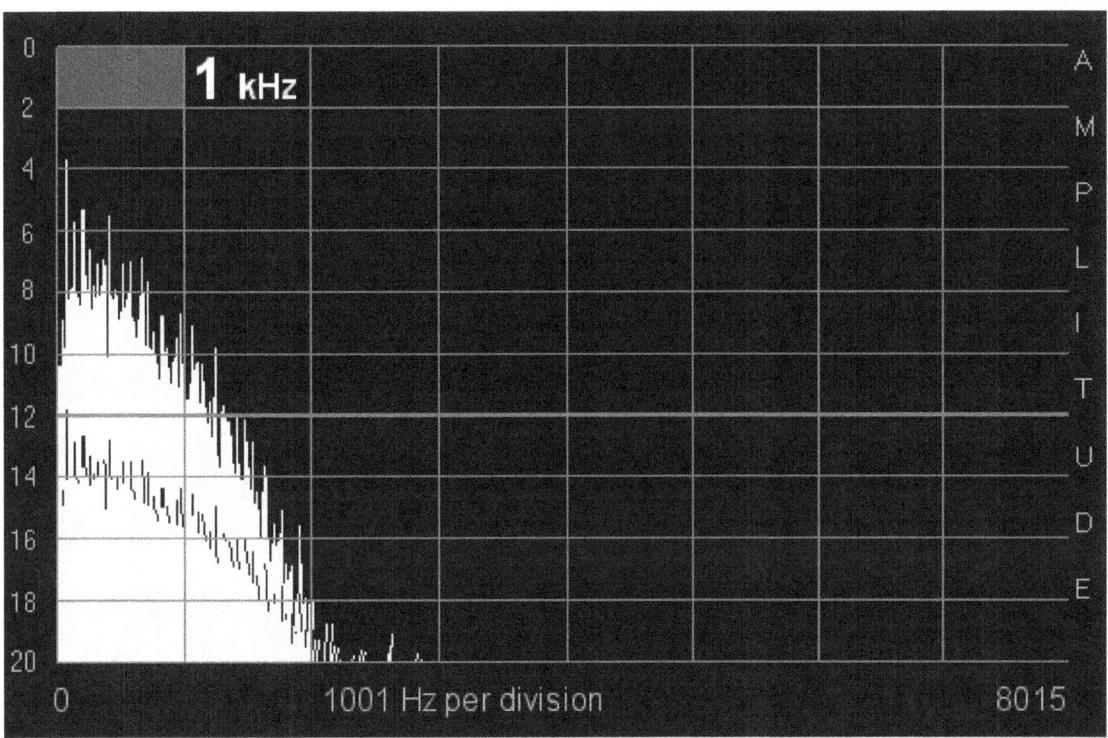

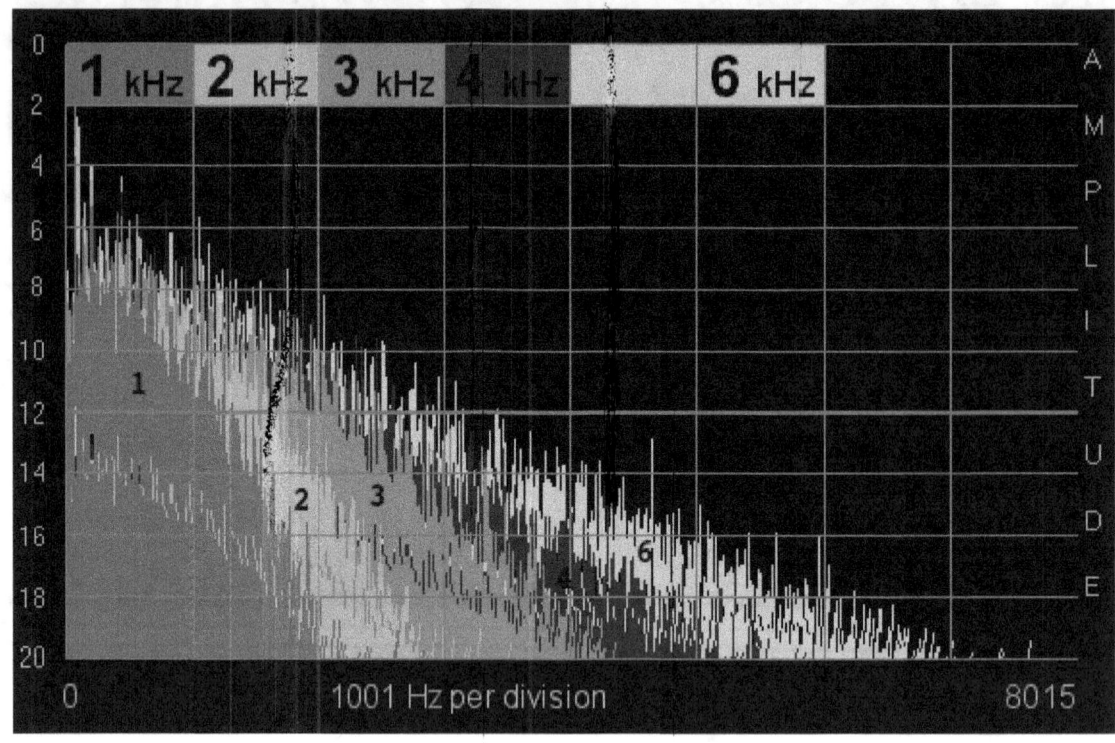

Speculating: the Si4734 hardware is 24-bit: this represents 144 dB of theoretical dynamic range. Each bit gives 6 dB. If the audio is 8-bit (takes 48 dB) then 96 dB of dynamic range is left.

3.7 Si4734 Offset Tuning
Maximum Offset Tuning Equals One Bandwidth
©2011

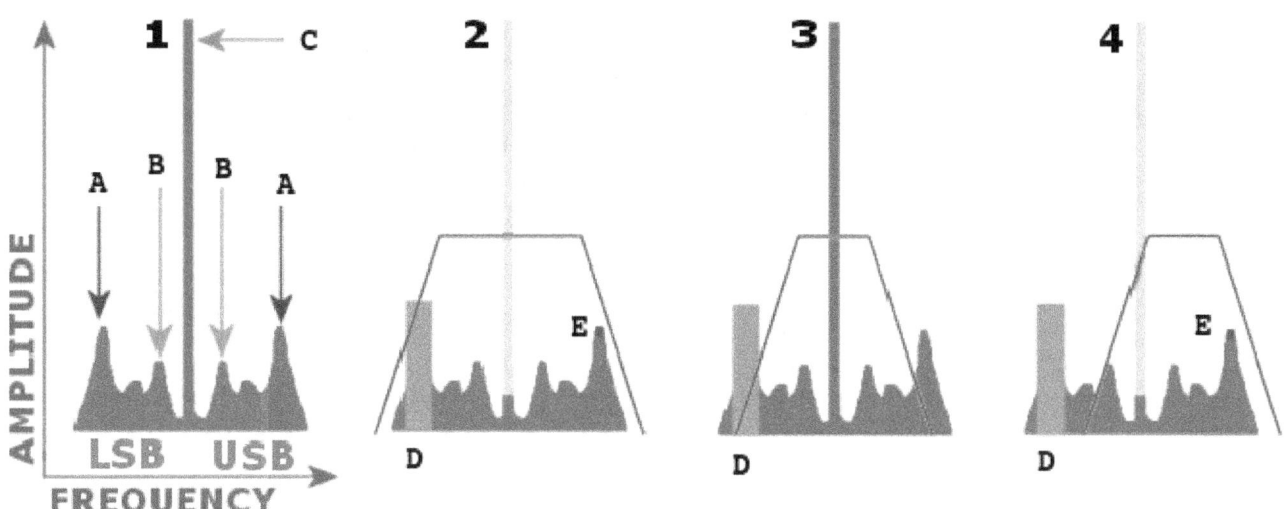

In **1** notice the high energy carrier (see "C") and positioning of the LSB and USB audio. Lower tones (see "B") are close to the carrier; while, higher tones (see "A") are far. In **2** a 6-kHz filter passes noises (see "D") in AM-mode. In **3** a 4-kHz filter drops fidelity and still passes some noise. In **4** upward (USB) offset-tuning of the 4-kHz filter attenuates the noise and boosts fidelity.

Compare a carrier dropout (selective fading) in **2** and **4**. In **2** the 6-kHz filter has 3 kHz of fidelity, the USB peak (see "E") can mix with noise, and up to 6 kHz noise can be emitted. In **4** the offset tuned 4-kHz filter has *nearly* 4-kHz of fidelity and noise of *only* 4-kHz may be emitted.

PL-390 Filter	Station Tuned	Distortion Noted	Maximum Offset Tuning	Ceramic Filter -6dB
6 kHz	6125 kHz	6132 kHz	6131 – 6125 = 6 kHz	12 kHz
4 kHz	6125 kHz	6130 kHz	6129 – 6125 = 4 kHz	8 kHz
3 kHz	6125 kHz	6129 kHz	6128 – 6125 = 3 kHz	6 kHz
2 kHz	6125 kHz	6128 kHz	6127 – 6125 = 2 kHz	4 kHz
1 kHz	6125 kHz	6127 kHz	6126 – 6125 = 1 kHz	2 kHz

The PL-390's audio was examined using a Fast Fourier Transform (FFT) program I created. The Si4734 DSP has filter bandwidths of 1, 2, 3, 4, and 6 kHz. They do not correspond with filters (*ceramic* or *mechanical*) in analog radios. They are different in two ways. First, they represent the width of audio emitted. To get 6 kHz of audio from a centered (around the carrier) analog filter, a ~12 kHz filter is needed. And this leads us to the second difference: DSP filters have a shape that may be very tight. Analog filters have a more gradual slope, as defined by -6 dB and -60 dB specs.

In the chart above, each filter was offset tuned until distortions were noted. The maximum offset equaled the filter bandwidth. A Si4734-based radio can be offset tuned by nearly the entire width of the filter. Some offset tuning, along with upping the volume, often helps with SWL and DX. Offset tuning up or down helps with adjacent interference and distortion related to selective fading.

3.8 Current Wars
Power Efficient Shortwave Radios
©2013

     I tested five portables I recommended (PL-390, DE1102, 7600GR, DE1103, KA2100); as well as the new **KA321** mechanically-tuned DSP radio and the not-recommended KK-9803 DSP radio. In the chart below, current was measured with no volume and not tuned to a station. Highlights include: 1) an increase in current when the KA321 uses earphones, 2) compared to the PL-390, the DE1103 consumes **2.94 times** the current and **3.92 times** the power, 3) compared to the PL-390, the 7600GR, DE1103, and KA2100 have inefficient backlighting, and 4) the KA2100 consumes 1.67 Watts while using its big speaker. The last column is power, on shortwave, in mW.

Radio	AA	FM (mA)	MW (mA)	SW (mA)	Ear (ΔmA)	Light (ΔmA)	SW (mW)
KA321	2	26.0	21.2	21.9	11.3	N/A	65.7
PL-390	3	22.7	20.5	24.9	0.0	6.1	112.1
KK-9803	2	40.4	40.2	40.3	0.0	N/A	120.9
DE1102	3	48.3	55.7	50.0	-3.9	8.3	225.0
7600GR	4	40.9	66.9	64.4	0.0	15.7	386.4
DE1103	4	57.6	73.6	73.2	0.0	19.6	439.2
KA2100	4	274.0	278.0	278.0	-203.0	12.0	1668.0

AA (number of AA batteries), Ear (earphone), Light (backlight), Δ (change in)

Silicon Labs

     The efficiency champions were the **PL-390** (Si4734) and **KA321** (Si4835). The, now, "old-school" 7600GR, DE1103, and KA2100 are pretty power hungry. The DE1103, for example, feeds many IC's: 2 mixer chips, a tuner chip, a PLL chip, 2 MCU chips, a memory chip, an audio amplifier chip, a line-out chip, and an FM front-end chip. The DE1103 also uses passive filters: bandpass, roofing, and ceramic IF. The truth is, the champion is not a radio, but a company: **Silicon Labs**. Their highly efficient CMOS, low-IF, DSP receiver chips are revolutionizing radio.

3.9 Silicon Labs Si4831/35
Mechanical Tuning AM/FM/SW Radio
©2011

Silicon Labs has just changed radio forever. The 24-pin SSOP Si4831 (AM/FM) and Si4835 (+SW) are DSP low-IF radio IC's, like the Si4734, but with an analog control interface. They allow mechanical tuning by reading a potentiometer using an analog-to-digital converter (ADC). Bands (5 AM, 5 FM, or 16 SW) are selected via *external resistances*. *Volume* (up/down in 32 levels) and *tone* (base/treble in 8 levels) controls are integrated. The tuner can indicate *station tuned* or *FM stereo* via lighting of external LEDs. An audio conditioner removes popping, clicks, and loud static.

Current consumption is 17 mA (10 µA in sleep mode) from, for example, 2 AAA batteries (2.0 to 3.6V). Coverage is 504 to 1750 kHz (AM), 64 to 109 MHz (FM), and 5.6 to 22.0 MHz (SW). Sensitivity on AM is specified at 25 µV for 26 dB SNR; this calculates out to 1.8 µV for 3 dB SNR.

This could mean the end of 455-kHz ceramic IF filters and switched LC tanks in lower-end analog radios. We will see very small DSP radios *without a noisy MCU, display, and its driver*; or their associated shielding. The units will offer *much better dial linearity* than the capacitor-tuned radios they replace. Below is an UNTESTED Si4831-based radio for MW hobbyists that *could easily fit inside of a match box*. **Silicon Labs** is rapidly becoming the new "Sony" of the radio industry.

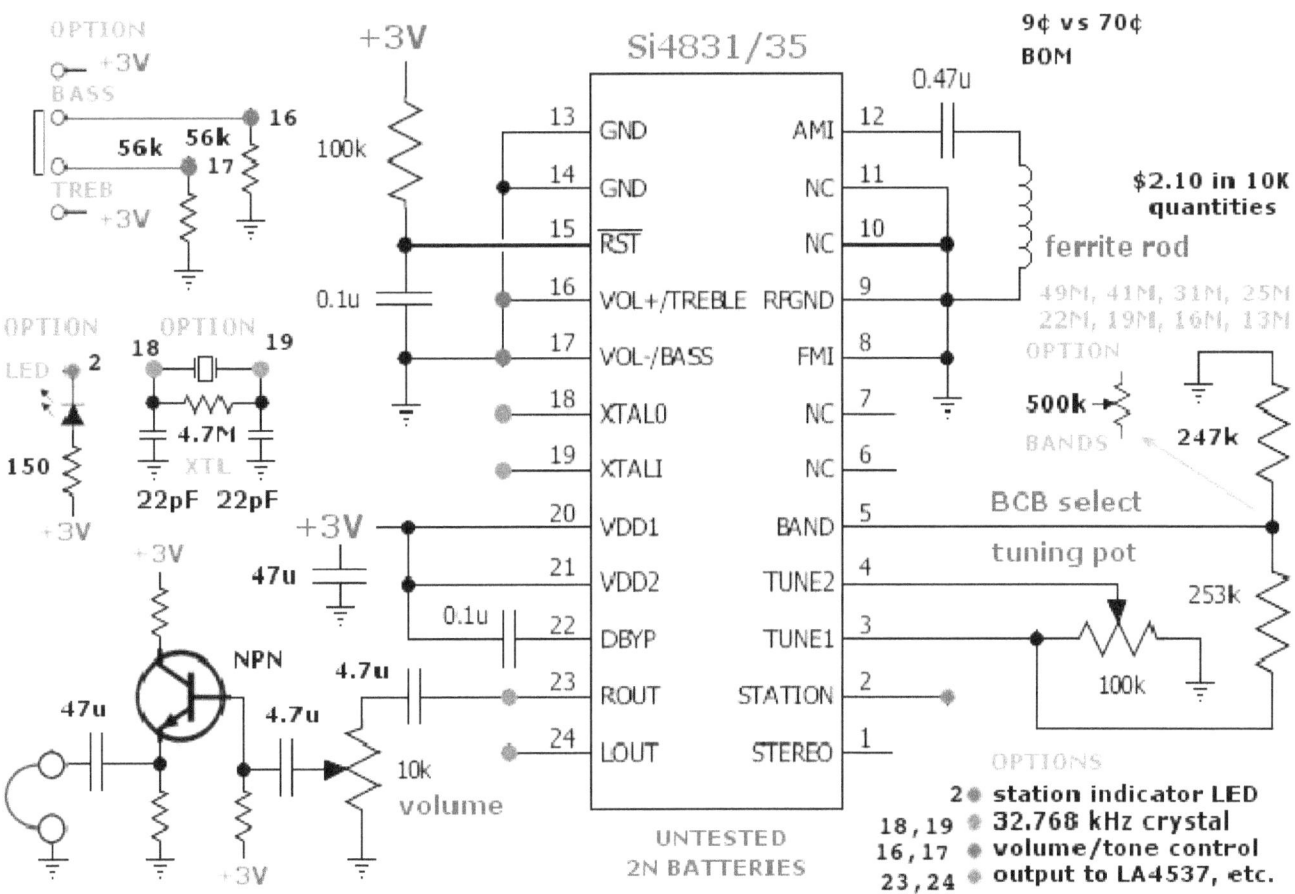

3.10 Hacking the KA321/Si4835
A million dollars of radio technology for $20.
©2013

1. INTRODUCTION

To the casual observer, the **KA321** looks like just another cheap analog portable: it's not. The KA321 contains **Silicon Labs**' potent Si4835 mechanically-tuned, low-IF, CMOS, FM/MW/SW, DSP, radio receiver chip. The KA321 makes those cheap, drifting, capacitor-tuned, ceramic-filtered, single-conversion, image-laden radios seem antiquated. Silicon Labs, of Austin, Texas, has solved the problems that plagued inexpensive, analog radios. The Si4835 combines: sensitivity (with an amplifier enable line for shortwave DX), selectivity using steep DSP filters (that ceramic filters cannot match), stability, blocking of images, blocking of adjacent interference, decent IP3, and linear tuning using a potentiometer instead of a variable capacitor (no hand capacitance effects).

The KA321 is small and light at 10.5 cubic inches and 162 grams (with batteries). Build is solid. The separate on-off switch is crisp. The internal speaker is surprisingly good. The volume control rotates easily. The tuning wheel has reduction drive (3.7:1). The band switch moves easily and its red indicator is easy to see. There is a red tuning LED; and a stereo headphone jack. The stand and battery door are adequate. The antenna is small at 18 inches but is rotatable. The radio comes with a lanyard loop. The KA321 can be powered through its 5 Volt DC power adapter jack.

Many aspects of the design are impressive; however, there is room for improvement. The current drain (on shortwave) using earphones is 11.3 mA **higher** than the 21.9 mA consumed by the internal speaker (see modification below). In low lighting, it is hard to read the shortwave band designations on the dial face (white lettering on a gray background). The font is small because "SW1" was written; instead of just "S1". US and European MW coverage would be ideal.

R3	=	47k
R4	=	20k
R5	=	120k
R6	=	39k
R7	=	20k
TOT	=	246k

2. MEDIUMWAVE COVERAGE

The KA321's dial reads 522 at the bottom and 1710 at the top. The problem here is that the band covered is either 520 to 1710 (AM1, US, 247k) or 522 to 1620 (AM2, European, 257k). In the PCB picture above we see that the AM band is set to 246k or the 520 to 1710 kHz US band.

MW coverage was confirmed before disassembly via a home designed frequency generator ($2 TLC555 based; 10-turn pot; tuned by listening for silence on a nearby digital radio). With the generator tuned to 1710 kHz, the KA321 could hear the silent "carrier", confirming its MW range.

3. FREQUENCY COVERAGE

The chart below was made using resistors in my KA321. The two FM bands are set to 75 µs de-emphasis. Shortwave is extended coverage. Resistance is total to ground; and sets each band.

KA321 Band	Band Name	Frequency Range	Resistance	Channels
FM1	FM1	87–108 MHz	67k	105
FM2	FM5	64-87 MHz	226k	115
AM	AM1	520–1710 kHz	246k	119
SW1	SW1	5.6–6.4 MHz	297k	160
SW2	SW3	6.8–7.6 MHz	317k	160
SW3	SW5	9.2–10.0 MHz	337k	160
SW4	SW7	11.45–12.25 MHz	357k	160
SW5	SW9	13.4–14.2 MHz	377k	160
SW6	SW11	15.0–15.9 MHz	397k	180
SW7	SW13	17.1–18.0 MHz	417k	180
SW8	SW15	21.2–22.0 MHz	437k	160

Nighttime shortwave (SW1,SW3,SW5); Daytime shortwave (SW7,SW9,SW11)

4. PRINTED CIRCUIT BOARD

The KA321's one-sided, surface-mount PCB consists of: 3 IC's, 4 transistors, 1 dual diode, 1 inductor, 26 capacitors, 35 resistors, and 9 zero-ohm jumpers. The component side includes: 12 jumpers, 9 electrolytic caps, 3 hole-through resistors, 2 pots, 1 32.768 kHz fork, 1 ferrite rod, 1 red LED, 1 power diode, 1 band switch, 1 on-off switch, 1 DC adapter jack, and 1 earphone jack.

The MW ferrite measures 1.93 inches (49mm x 8mm x 5mm) and is covered by ~92 turns. Care must be taken not to damage its two thin leads. However, adding a new ferrite is safe as both the AM input and ground input are sturdy jumper lines on the component side of the board.

YAG laser markings on IC1 read: 4835B30GU, with a date code of week 26 of 2011. The KA321 uses Silicon Labs' second generation chip. IC2 reads: CD1622CB, a stereo/monaural BTL power amplifier. IC3 helps feed IC1 a steady 2.13 Volts. Two transistors are used as RF amplifiers (SW hfe=267, FM hfe=487; NPN; both connected common-emitter); the other two are switches.

The printed circuit board is impressive but looks like it had a quick wash and not enough dry time before applying the conformal coating (caked white residue). There is also solder balling.

5. DISASSEMBLY

The KA321 is disassembled by removing: a) 4 back and 1 antenna screws, b) 2 speaker, 2 battery, and 1 antenna wires, c) a PCB screw, d) 4 tuning mechanism screws, e) 2 runs of glue under the dial face's left side plus tiny dabs on the right side and top, f) a tuning wheel and its retainer, g) a band switch extension, h) a pot wheel and its screw, i) a dial pointer, its cord, and its load spring, j) a red LED, and k) glue that holds the plastic tuning housing to the MW ferrite.

6. DIGITAL FILTERS

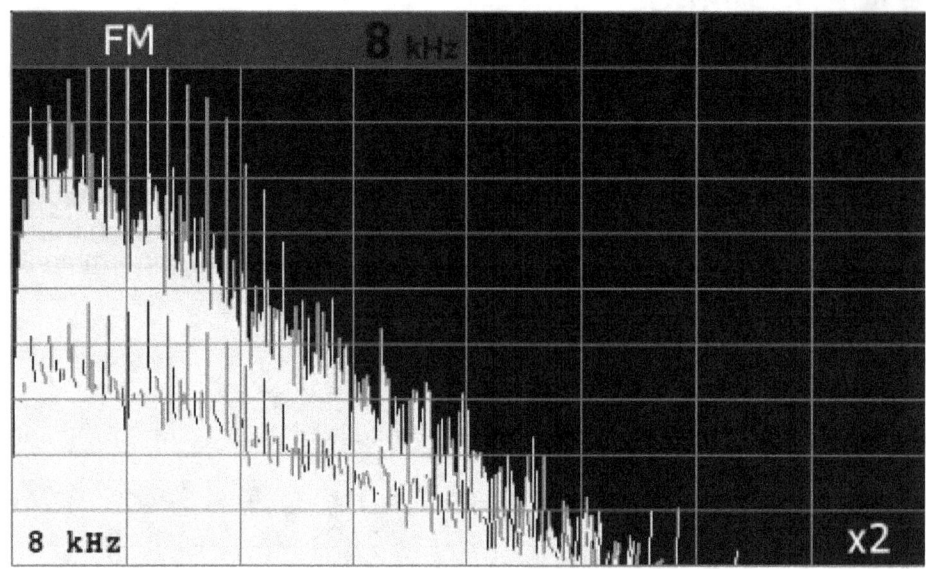

The graph above shows the attenuation of background noise via the KA321's FM filter. The sample rate is 48,000 Hz. The horizontal axis is linear: 2003 Hz per division. The vertical axis has a log 10 scale: 2 dB per division. The trace represent peeks over a timed period. This is an 8-kHz (audio) filter. Compare this to the known 4-kHz filter in the PL-390 below (horizontal axis: 1001 Hz per division). The 3-kHz filter and 2-kHz filter of a PL-390 are also shown below for comparison purposes. Shape factors for the Si4835's DSP filters are tighter than even quality, ceramic filters.

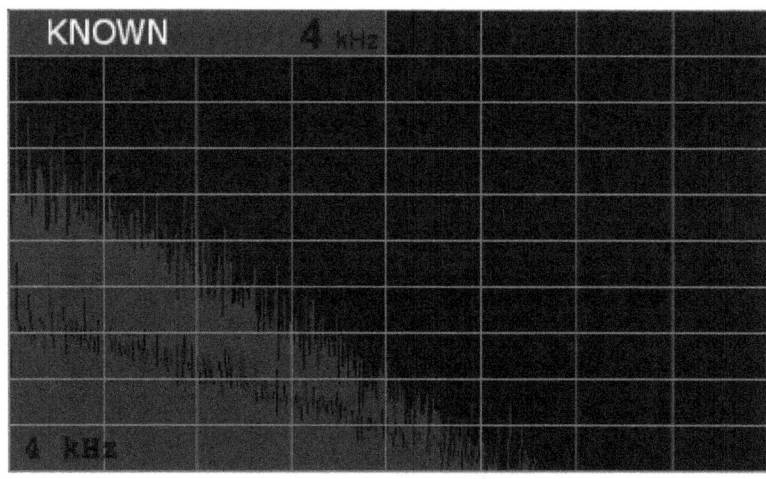

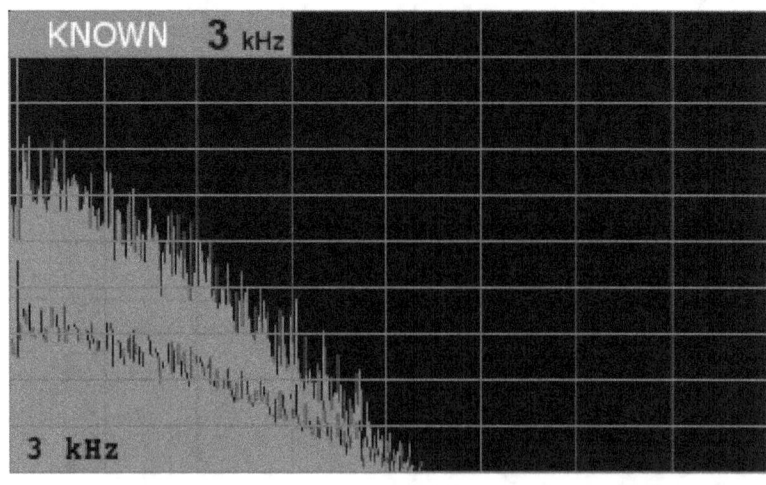

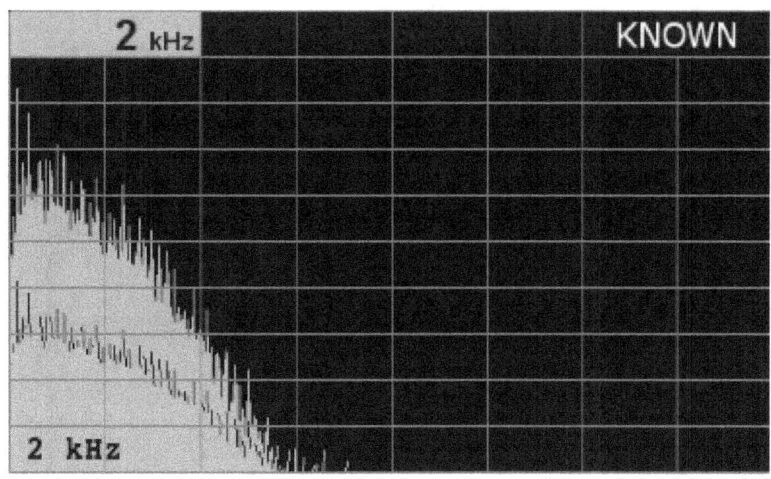

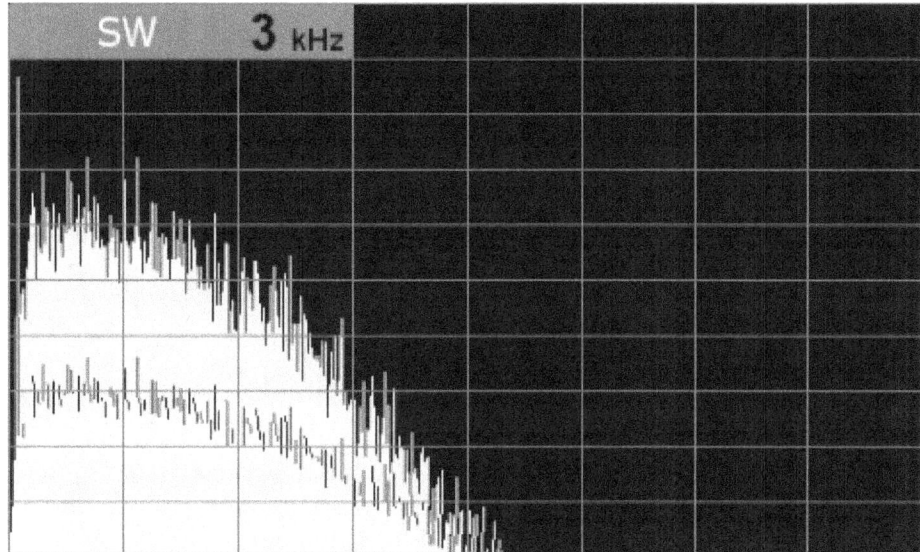

The graph above shows the attenuation of background noise via the KA321's SW filter. The sample rate is 48,000 Hz. The horizontal axis is linear: 1001 Hz per division. The vertical axis has a log 10 scale: 2 dB per division. The trace represent peeks over a timed period. This is a 3-kHz (audio) filter. Compare this to the known 3-kHz filter in a PL-390 (above). To achieve 3-kHz of audio in a non-DSP radio would require centering a 6-kHz filter (ex. ceramic) over an AM signal.

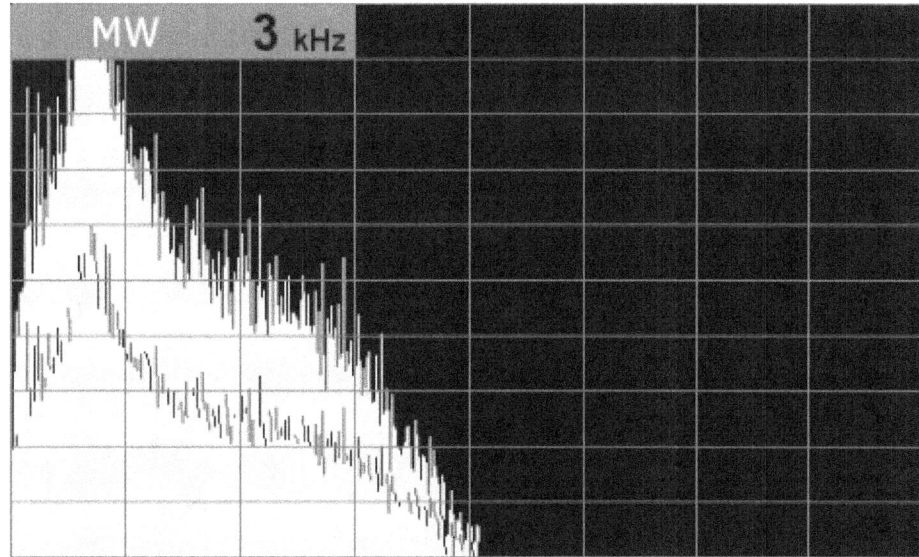

The graph above shows the KA321's MW filter. (same sample rate, scale, and timing period). This is also a 3-kHz (audio) filter. Silicon Labs' FM, SW, and MW filters are all well chosen.

7. UNOFFICIAL KA321 BLOCK DIAGRAM

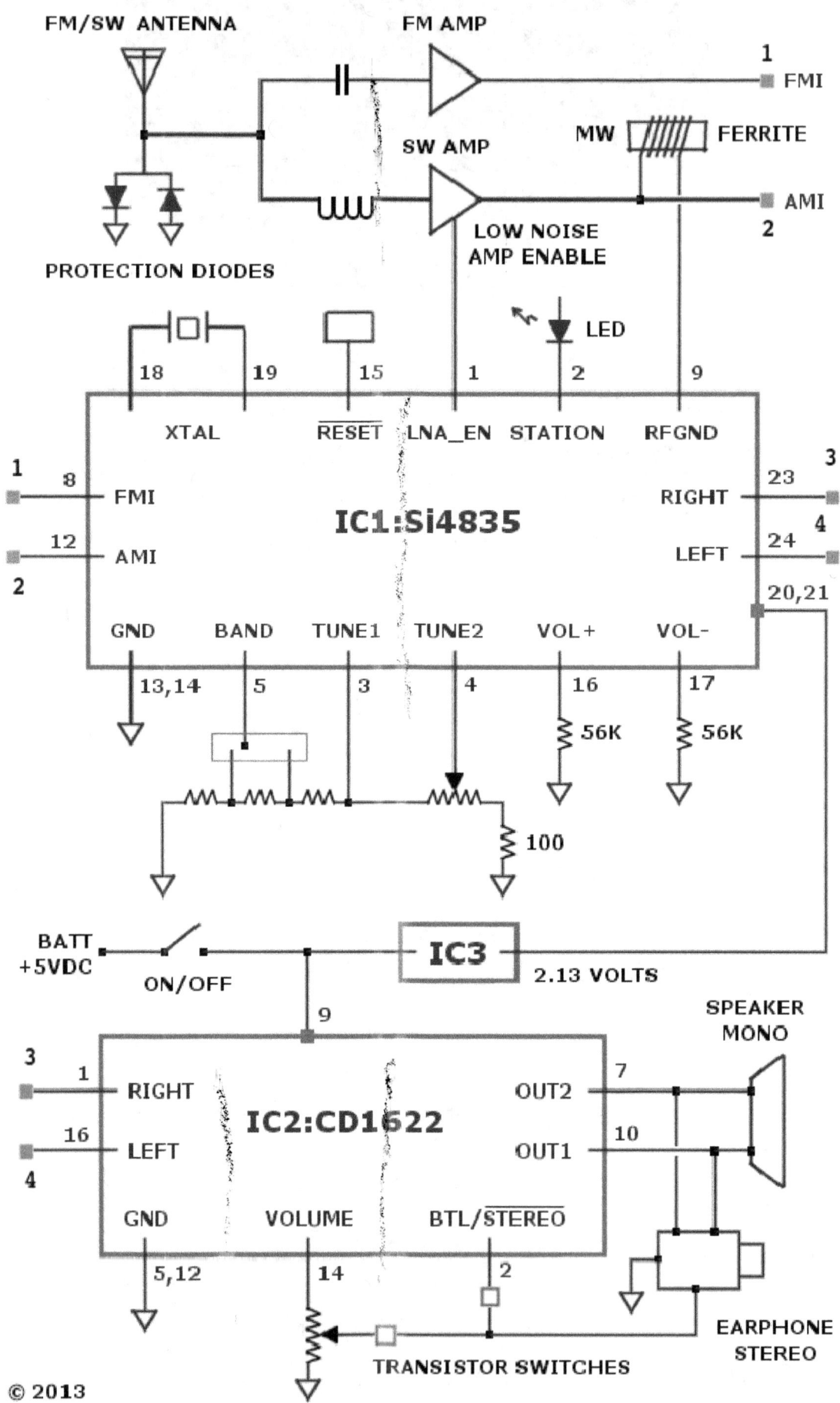

8. TONE CONTROL MODIFICATION

Adding a 3-level base-normal-treble control was too simple to pass up. The 56k resistors to ground are already present on *pin 16* and *pin 17* [this sets the volume to maximum]. And there is no 10k pull-up resistor on *pin 2* so base-treble mode is selected [versus volume mode]. Attaching Vcc to *pin 16* creates bass; to *pin 17* creates treble (some documentation discrepancies exist). The Si4835 also supports a 9-level base-treble control. Silicon Labs may wish to add bass boost. Select a radio with well-liked bass boost and either run tones through its filter or PSpice its schematic. The 1800 Hz bass filter is good for DXing and could easily have been added by Degen.

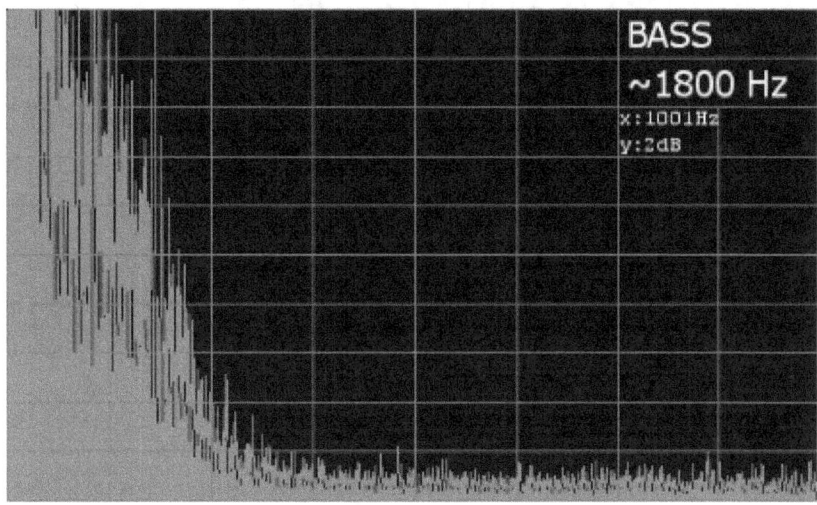

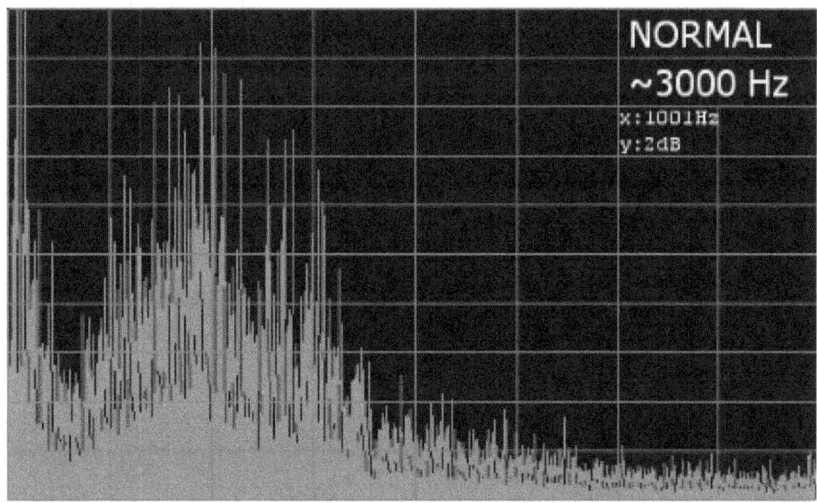

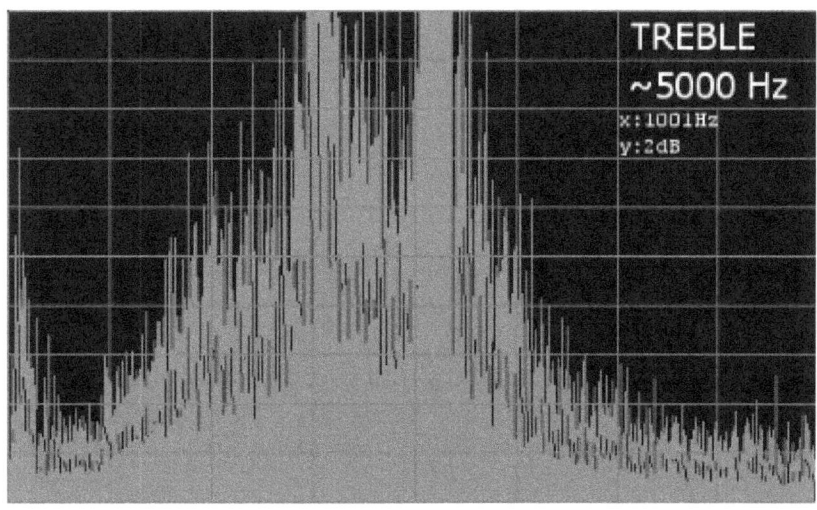

9. HEADPHONE OUTPUT MODIFICATION

The headphone output design needs modification. Output capacitors should not be omitted in stereo mode, as they are in mono bridge-tied-load mode. Each output needs a 100 ohm resistor; followed by a 47 µF capacitor. The resistor would sink some energy, allowing Degen to do away with the volume control acrobatics. The capacitors will drop stereo headphone current consumption from 33.2 mA to 21.9 mA; a 51% decrease in power use. This can be done outside the radio by adding 47 µF electrolytic caps to the left and right line; positive side toward the chip.

10. TUNING POTENTIOMETER MODIFICATION

A 25-turn, 100k-ohm, cermet potentiometer was added to tune the KA321. This involved altering the radio: using snipers to shatter, like glass, R1 [a 100 ohm SMD resistor]. The shards were removed and the new pot tapped into two of the original pot's lines and into ground (via a 100 ohm resistor). The Si4835 can easily be heard transitioning to each new frequency channel.

11. ANTENNA NOTES

The KA321 is so sensitive that directly connecting a longwire antenna can overload it. Just bringing a 25 foot wall-mounted wire close to the KA321's antenna increases the signal. It can be physically attached but through a very low capacitance feed, such as 1 to 22 pF. Although the KA321 has a small MW ferrite, inductive coupling with a tuned MW loop yielded many AM stations.

12. DUAL D BATTERIES

Adding Duracell D batteries (15,000 mAh) allows operating the KA321 on shortwave for up to 685 hours. The Si4835, itself, uses 17 mA on SW; the KA321's design adds only 4.9 mA more.

13. KA321 VERSUS PL-390

I recommend the PL-390 in my latest shortwave guide. A casual comparison of the KA321 (chart below) revealed that it was as capable at detecting stations as the PL-390. However, it took finesse to tune many of the stations. This was before the addition of the 25-turn, cermet potentiometer. Note that additional wire was capacitatively coupled to the KA321, as needed. The KA321 engineers did a great job with the off-chip RF section of the radio. Initially, on FM, while DXing, only 37 of 42 stations were heard. However, each was found later by going back and holding the 18 inch KA321 antenna higher by 3 feet. The KA321's RF performance is impressive.

Test	Time	Band	PL-390 Stations Heard	KA321 Stations Heard	KA321 %
ETM	8 PM	FM	28	28	100
ETM	8 PM	SW	24	24	100
ETM	11 PM	FM	26	26	100
ETM	11 PM	SW	17	17	100
DX	3 PM	25M	10	10	100
DX	3 PM	22M	4	3	75
DX	3 PM	19M	6	6	100
DX	7 PM	49M	14	14	100
DX	7 PM	41M	7	7	100
DX	8 PM	49M	18	18	100
DX	NIGHT	49M	17	17	100
DX	NIGHT	FM	42	42	100

Electromagnetic energy obeys the laws of physics and does not respond according to how much we like or spend on a radio. In 2002, I wrote an article (see: www.radiointel.com) that showed that an $11 Sony S10MK2 with a $20 Radio Shack loop could hear 94% of what an ICOM IC-R75 with a Quantum QX loop ($780 total cost) could hear. That page has over 28,000 views.

14. FREQUENCY DETERMINATION

The Si4835 contains a low-dropout [LDO] voltage regulator and is fed a steady 2.13 Volts. The chip reads the position of the 100k tuning potentiometer via an ADC or analog-to-digital converter. An ADC reads voltage. Radio hobbyists can read voltage, too. Harbor Freight sells a multimeter for $8, often on sale for $5; although, I got mine free with a coupon. The black lead was connected to ground; the red lead to *pin 4,* aka TUNE2, aka the potentiometer wiper. Using a 2000 mV range, voltages were read, continuously, while tuning. There is a linear relationship between tuning dial position and the frequency tuned. Therefore, there is a linear relationship between potentiometer wiper resistance, frequency tuned, and the voltage read. Below is a plot of tuned frequency in MHz (x-axis) [FM1 band] versus read meter voltage in mV (y-axis); it is linear.

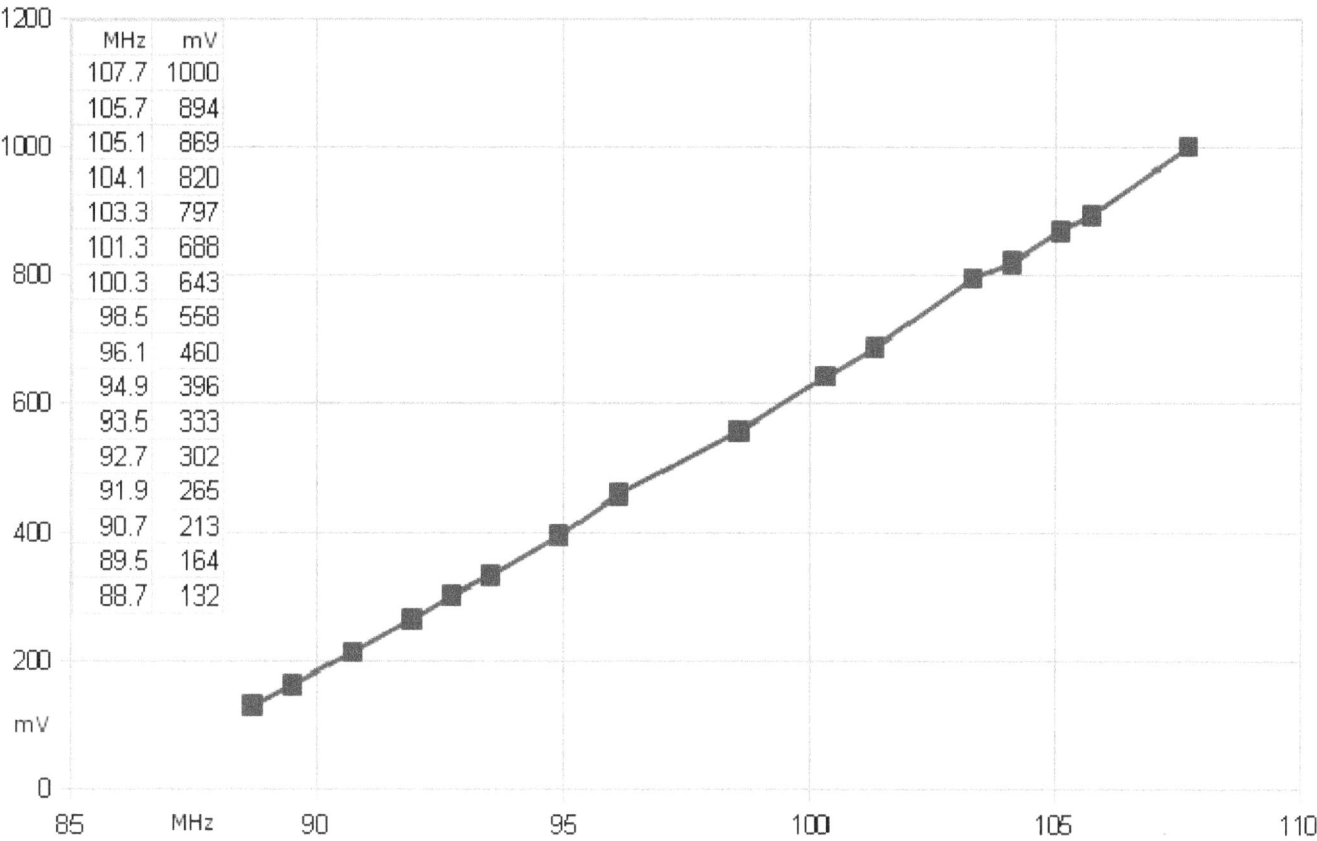

MHz	mV
107.7	1000
105.7	894
105.1	869
104.1	820
103.3	797
101.3	688
100.3	643
98.5	558
96.1	460
94.9	396
93.5	333
92.7	302
91.9	265
90.7	213
89.5	164
88.7	132

Using a digital spotter radio or by listening for station identification, a chart of voltage versus frequency can be created, for each band. It is possible to operate the KA321 radio and know where it is tuned or tune it to a frequency. This aids DX work. Of Silicon Labs' chips, I like the Si4835 best. When a radio manufacturer adds a microprocessor, display driver, and LCD; they also add noise. And, on portables, shielding is often an afterthought. Would you put an analog radio inside a running computer's case? Yet manufacturers will stick an MCU centimeters from the RF energy. The Si4835 contains an MCU, but uses a patented ratiometric clock, to counter digital harmonics. One oscillator is divided down for digital clocking and for local oscillator generation.

15. HELLENIZED FREQUENCY INDICATOR

The Si4835 engineers allow designers to create a superb radio without: an RF input filter, a front-end mixer, a roofing filter, a separate tuner chip, ceramic IF filters, a MCU, a memory chip, a PLL chip, an FM front-end chip, and a tuning capacitor. However, the KA321 design still strikes me as complex, due to its intricate dial mechanism. I envisioned a solution: use a voltage buffer off *pin 4* (TUNE2) connected to a mechanical ammeter with the dial printed on it. The ammeter would be setup to measure voltages from 0 to 1000 mV (1 Volt) with subsequent deflection of its needle pointer. The voltage buffer might be an op-amp based, unity gain, buffer amplifier. The buffer needs a high input impedance so that it will not disturb the Si4835's tuning mechanism.

16. ATTACK OF THE CLONES

The KK-9803 is a DSP radio that could be mistaken to contain an Si4835. Inside is a COB assembly with an IC labeled: DSP6919ZZ. The assembly also contains 1 inductor, 10 capacitors, 6 resistors, and diode protection [all SMD]. The DSP6919ZZ contains an audio amplifier, receives TV audio, and is tuned via a 40 pF capacitor. Current use is 29 mA; powered-down to 33 µA [Si4835 to 10 µA]. It uses a 32.768 kHz fork and has 6 digital input lines to select between 4 FM, 1 MW, and 12 SW ranges. Below are graphs of the 4-kHz DSP filter used on MW and SW on the KK-9803.

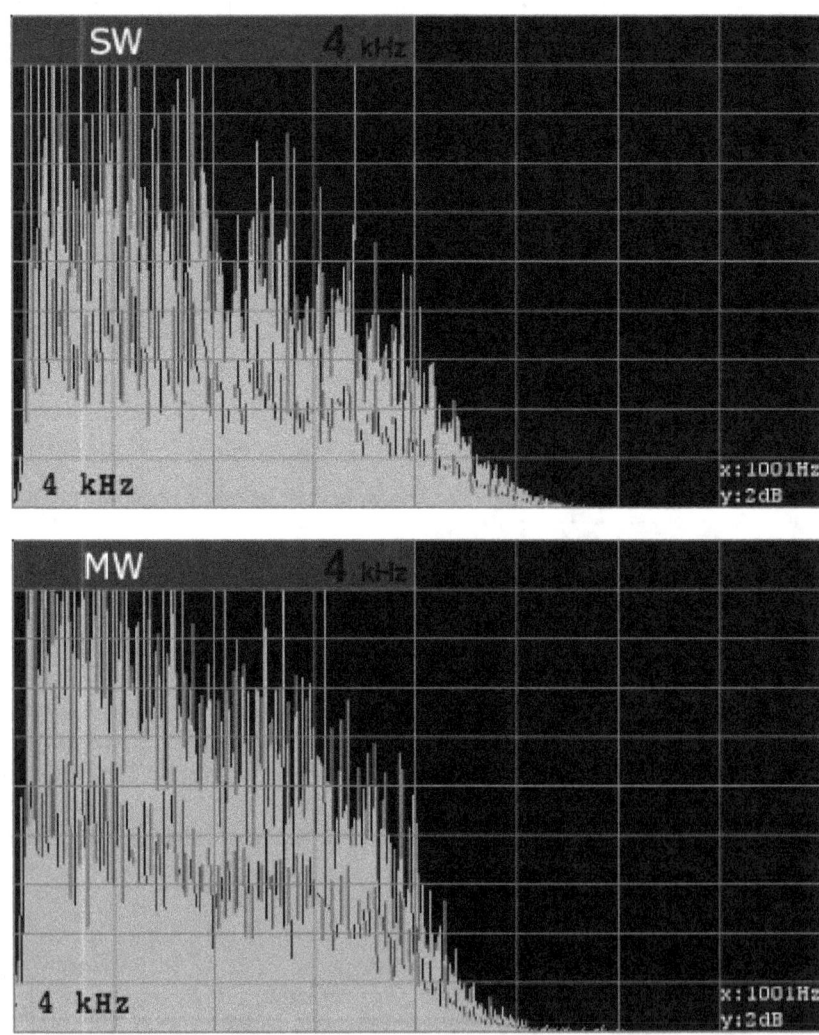

Another clone chip, the DSP6935, is similar to the Si4734 [Note: Version D60 now has 7 bandwidths: 6, 4, 3, 2, 1, 2.5, and 1.8 kHz]. The DSP6935 has an I2C interface, an audio amplifier, receives TV audio, and uses 60 mA. It allows the setting of high or low-side injection. Power down current is 470 µA (sic) [Si4734 is 9.5 µA]. Is this a case for "*Silicon Labs Inside*" stickers?

17. DISCUSSION

Silicon Labs' engineers have changed radio. And Kaito's KA321 (aka Degen's DE321) is the beginning of the end of analog radios that are capacitor-tuned, ceramic-filtered, and single-conversion. These radios do not have the linear tuning, filter shape, image rejection, or stability that Si4835-based radios offer. The Si4835 requires no manual alignment, so it is convenient for manufacturers, too. All this, in a chip that is so small that 8 of them could fit on a postage stamp.

Sony's own state-of-the-art CXA1129N, is an impressive, Hartley image rejection, low-IF, bipolar IC. It uses 15.8 mA at only 1.0V [Si4835 uses 17.0 mA at 2.0V]. However it is still tuned via a capacitor. Due to Silicon Labs, it's starting to look dated, like the pinnacle of a fading era.

Buy a KA321, it's a sleeper. It's the Dodge Dart of radios; with Silicon Labs' 440 6-pack under the hood and a pair of Dr. Phil's headers in the trunk. This, in a day when so many are offering Ford Mustangs with a V6 (digital frequency displays mated to single-conversion junkers).

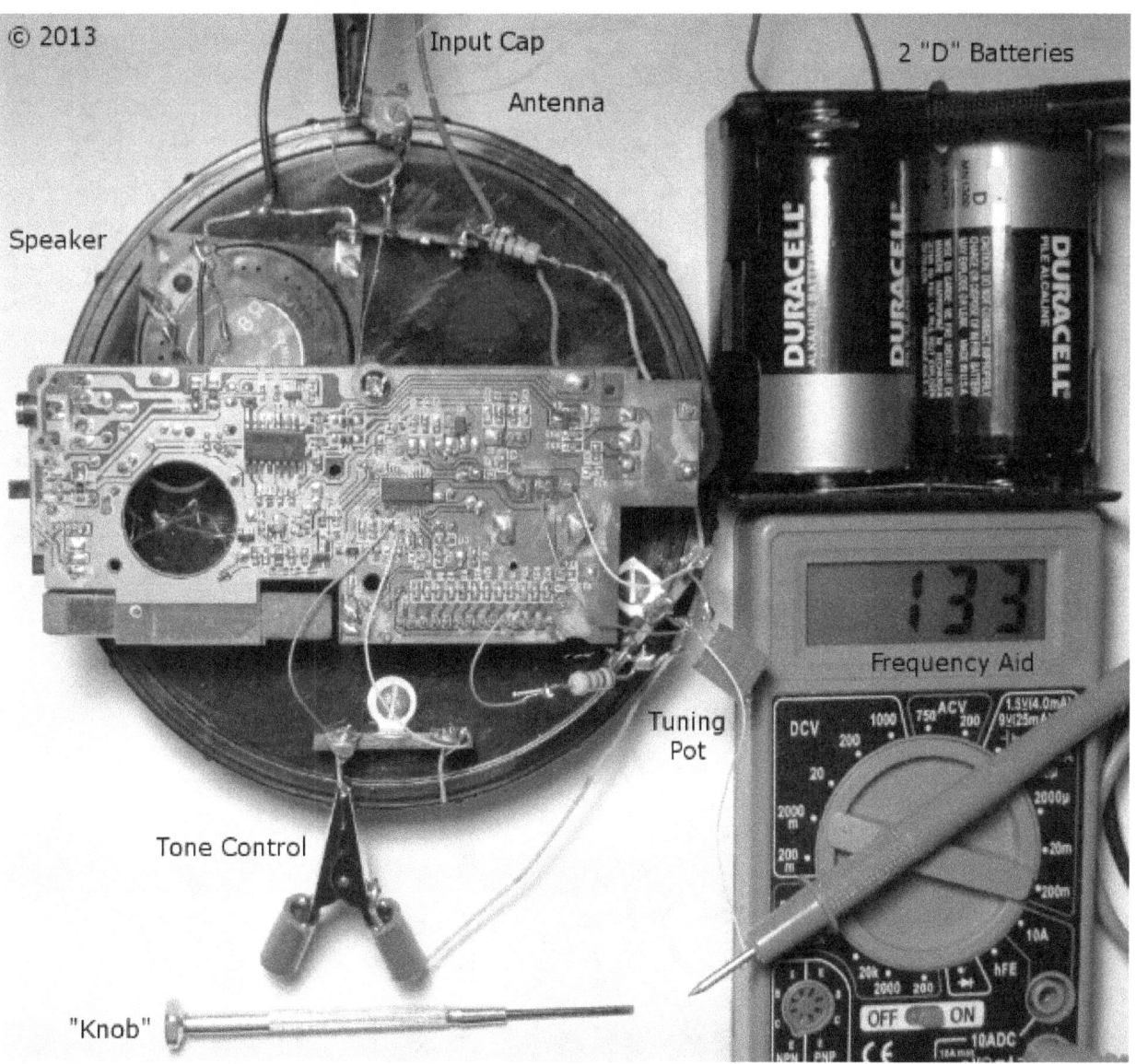

3.11 KA321 Update 2
10-turn pot and 2 AA with power switch
©2015

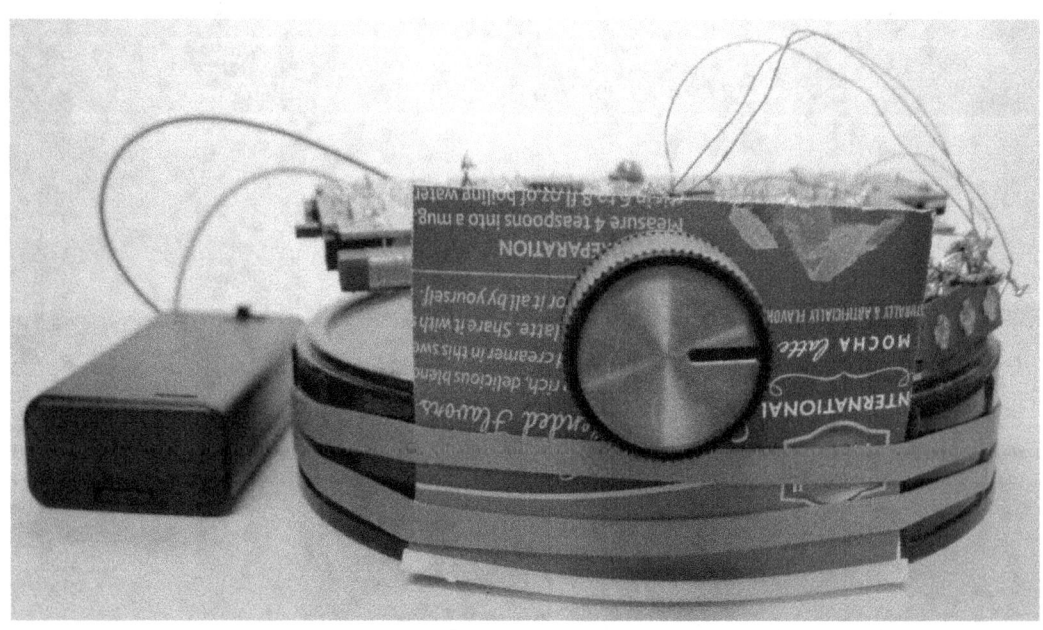

3.12 KA321 Voltage to Frequency: MW
Linear regression.
©2013

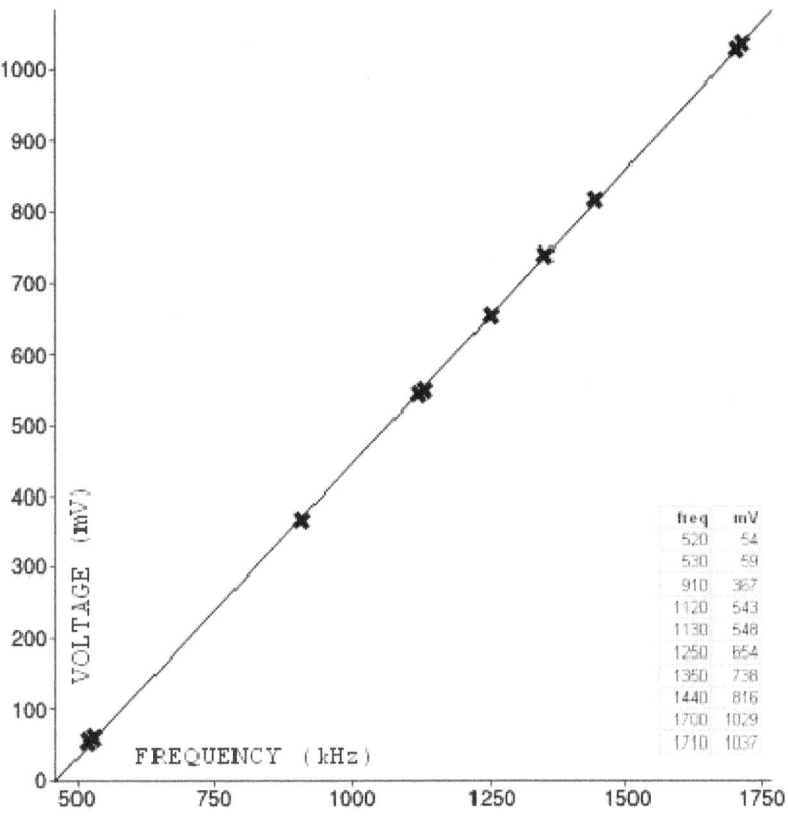

freq	mV
520	54
530	59
910	367
1120	543
1130	548
1250	654
1350	738
1440	816
1700	1029
1710	1037

Using a KA321, 4 local AM stations and 6 man-made carriers (520, 530, 1120, 1130, 1700, and 1710 kHz) were tuned on AM1. The voltage from ground to *pin 4* (TUNE2) was noted over the range where the tuning LED was lit. The average of this range, in mV, was plotted against frequency tuned, in kHz. The results are linear (see graph above). Linear regression established the relationship: **kHz = (mV + 382.51) / 0.82934**. The correlation coefficient, r, is 0.999937. The chart below correlates a mV reading at *pin 4* with a kHz frequency tuned. This will aid MW DX.

~mV	kHz	~mV	kHz	~mV	kHz	~mV	kHz	~mV	kHz	~mV	kHz	~mV	kHz	~mV	kHz	~mV	kHz
49	520	165	660	281	800	397	940	513	1080	629	1220	745	1360	861	1500	978	1640
57	530	173	670	289	810	405	950	521	1090	638	1230	754	1370	870	1510	986	1650
65	540	181	680	298	820	414	960	530	1100	646	1240	762	1380	878	1520	994	1660
74	550	190	690	306	830	422	970	538	1110	654	1250	770	1390	886	1530	1002	1670
82	560	198	700	314	840	430	980	546	1120	662	1260	779	1400	895	1540	1011	1680
90	570	206	710	322	850	439	990	555	1130	671	1270	787	1410	903	1550	1019	1690
99	580	215	720	331	860	447	1000	563	1140	679	1280	795	1420	911	1560	1027	1700
107	590	223	730	339	870	455	1010	571	1150	687	1290	803	1430	920	1570	1036	1710
115	600	231	740	347	880	463	1020	580	1160	696	1300	812	1440	928	1580		
123	610	239	750	356	890	472	1030	588	1170	704	1310	820	1450	936	1590		
132	620	248	760	364	900	480	1040	596	1180	712	1320	828	1460	944	1600		
140	630	256	770	372	910	488	1050	604	1190	721	1330	837	1470	953	1610		
148	640	264	780	380	920	497	1060	613	1200	729	1340	845	1480	961	1620		
157	650	273	790	389	930	505	1070	621	1210	737	1350	853	1490	969	1630		

3.13 KA321 Voltage to Frequency: SW
Linear regression.
VERSION 1 ©2013

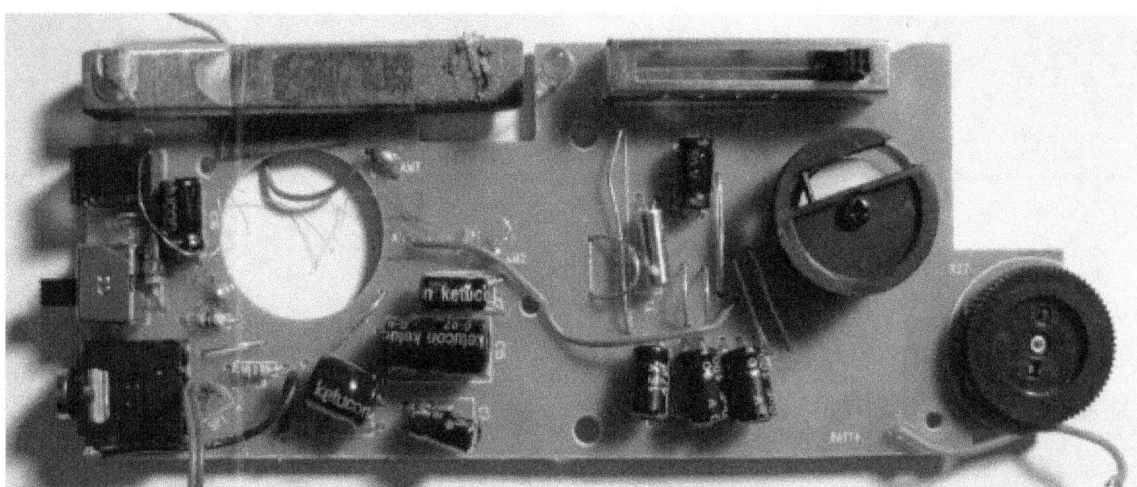

Using a KA321, five SW stations were tuned on six bands. The voltage from ground to *pin 4* (TUNE2; tuning pot wiper) was recorded over the range in which the tuning LED was lit. The average of this range, in mV, was plotted against frequency tuned, in kHz. The results are linear. Linear regression was run and equations relating kHz to mV were calculated. Equations allow a mV reading at *pin 4* to be converted into a frequency. The correlation coefficient, r, shows the linearity of the data. This research was done to aid DX work on shortwave using the Kaito KA321.

Band	Range	Equation	"r"
SW1	5600– 6400 kHz	kHz = 5572.77 + 0.796582 * mV	0.999987
SW2	6800– 7600 kHz	kHz = 6773.71 + 0.795289 * mV	0.999990
SW3	9200–10000 kHz	kHz = 9162.61 + 0.811359 * mV	0.999908
SW4	11450–12250 kHz	kHz = 11424.2 + 0.793609 * mV	0.999993
SW5	13400–14200 kHz	kHz = 13372.1 + 0.796723 * mV	0.999985
SW6	15000–15900 kHz	kHz = 14963.5 + 0.906189 * mV	0.999972

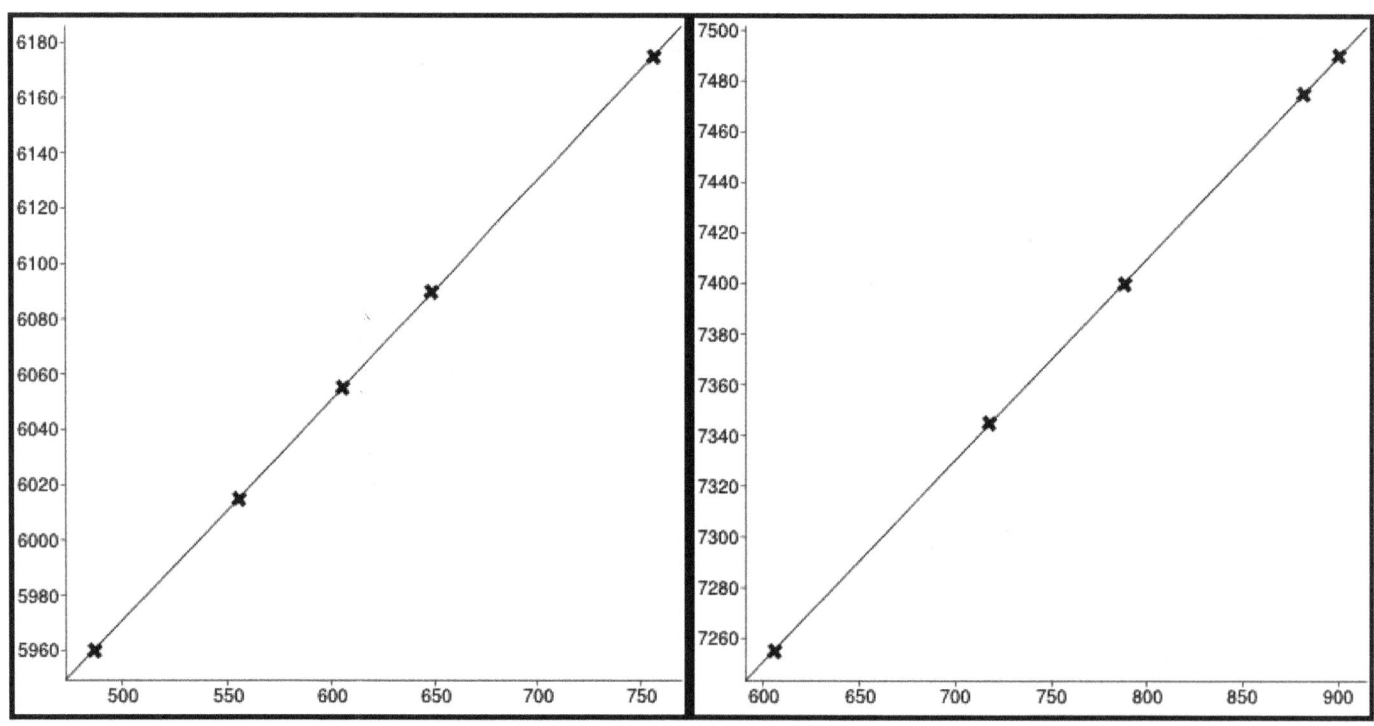

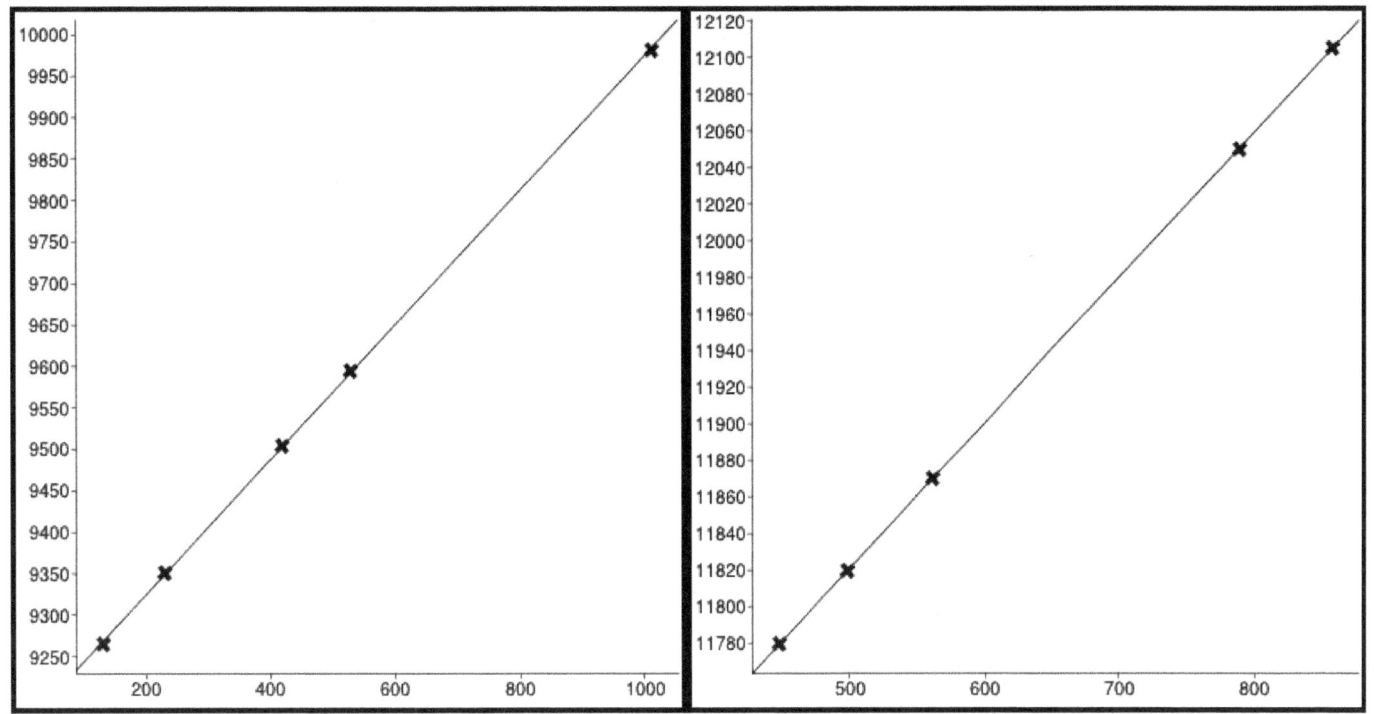

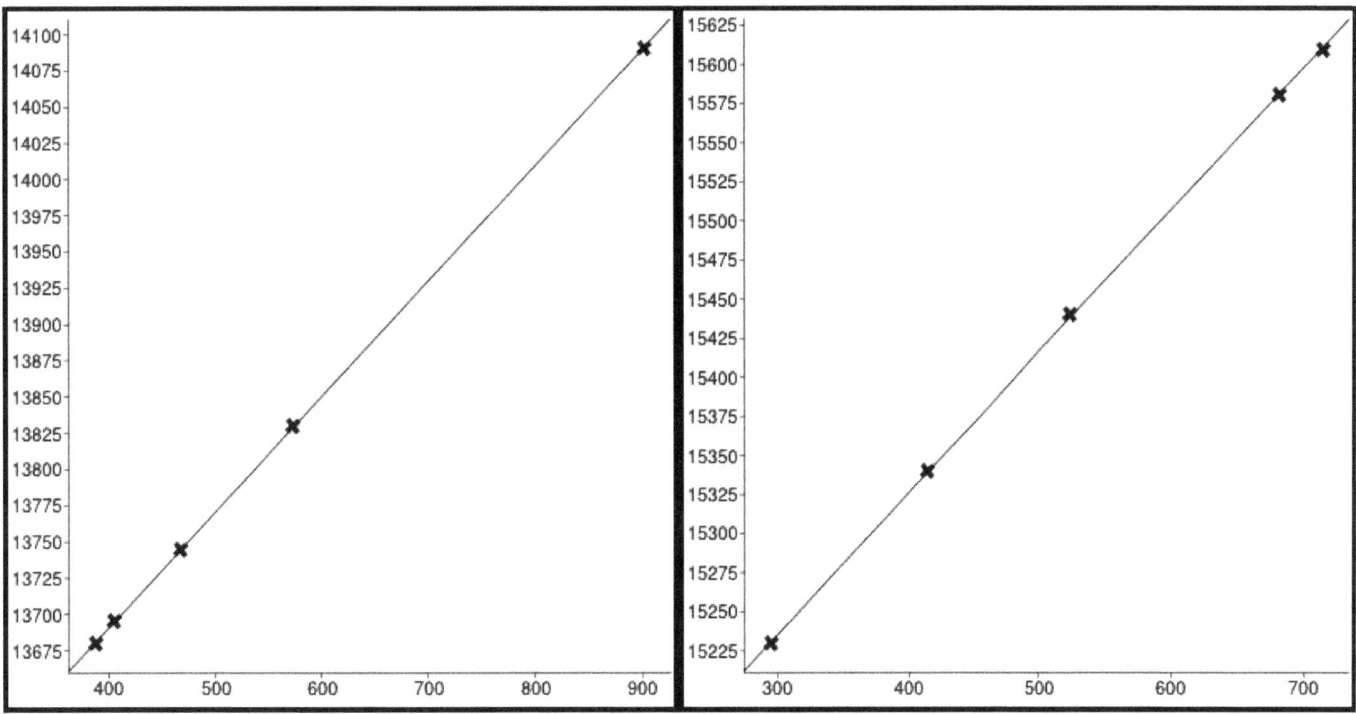

In the graphs above, the x-axis (horizontal) represents the reading at *pin 4* in mV; whereas, the y-axis (vertical) represents the frequency tuned, in kHz. Larger voltage ranges and more data points would have produced better results; however, due to the high correlation coefficients, the equations are likely accurate. It will be noted that *pin 3* aka TUNE1 should be thought of as a *reference voltage* against which the ADC compares *pin 4* aka TUNE2 or *pin 5* aka BAND voltages against, in relationship to ground. The ADC will feed this data to an internal MCU.

3.14 KA321 Voltage to Frequency: FM
Linear regression.
©2013

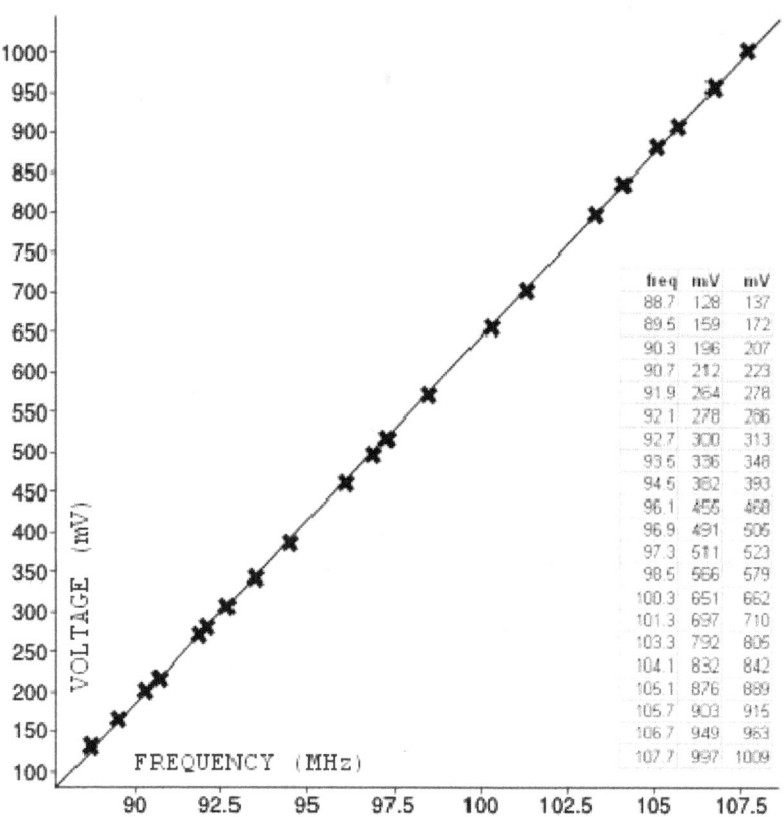

freq	mV	mV
88.7	128	137
89.5	159	172
90.3	196	207
90.7	212	223
91.9	264	278
92.1	278	286
92.7	300	313
93.5	336	348
94.5	382	393
96.1	455	468
96.9	491	505
97.3	511	523
98.5	566	579
100.3	651	662
101.3	697	710
103.3	792	805
104.1	832	842
105.1	876	889
105.7	903	915
106.7	949	963
107.7	997	1009

Using a KA321, twenty one local FM stations were tuned on the FM1 band. The voltage from ground to *pin 4* (TUNE2; tuning pot wiper) was recorded over the range in which the tuning LED indicator was lit. The average of this range, in mV, was plotted against frequency tuned, in MHz. The results were very linear (see graph above). Linear regression was run and an equation relating MHz to mV was calculated. This equation is: **MHz = (mV + 3963) / 46.081**. The correlation coefficient, r, is 0.999929. The chart below allows a mV reading at *pin 4* (the tuning potentiometer's wiper) to be turned into a frequency value on the FM band. This will aid FM DX.

~mV	MHz	~mV	MHz	~mV	MHz	~mV	MHz	~mV	MHz	~mV	MHz	~mV	MHz	~mV	MHz	~mV	MHz
51	87.1	161	89.5	272	91.9	382	94.3	493	96.7	604	99.1	714	101.5	825	103.9	935	106.3
60	87.3	171	89.7	281	92.1	392	94.5	502	96.9	613	99.3	723	101.7	834	104.1	945	106.5
69	87.5	180	89.9	290	92.3	401	94.7	512	97.1	622	99.5	733	101.9	843	104.3	954	106.7
78	87.7	189	90.1	300	92.5	410	94.9	521	97.3	631	99.7	742	102.1	853	104.5	963	106.9
88	87.9	198	90.3	309	92.7	419	95.1	530	97.5	641	99.9	751	102.3	862	104.7	972	107.1
97	88.1	207	90.5	318	92.9	429	95.3	539	97.7	650	100.1	760	102.5	871	104.9	982	107.3
106	88.3	217	90.7	327	93.1	438	95.5	548	97.9	659	100.3	770	102.7	880	105.1	991	107.5
115	88.5	226	90.9	336	93.3	447	95.7	558	98.1	668	100.5	779	102.9	889	105.3	1000	107.7
124	88.7	235	91.1	346	93.5	456	95.9	567	98.3	677	100.7	788	103.1	899	105.5	1009	107.9
134	88.9	244	91.3	355	93.7	465	96.1	576	98.5	687	100.9	797	103.3	908	105.7		
143	89.1	253	91.5	364	93.9	475	96.3	585	98.7	696	101.1	806	103.5	917	105.9		
152	89.3	263	91.7	373	94.1	484	96.5	594	98.9	705	101.3	816	103.7	926	106.1		

3.15 PL-390: FM DX King
This DSP-based radio shines on FM.
©2011

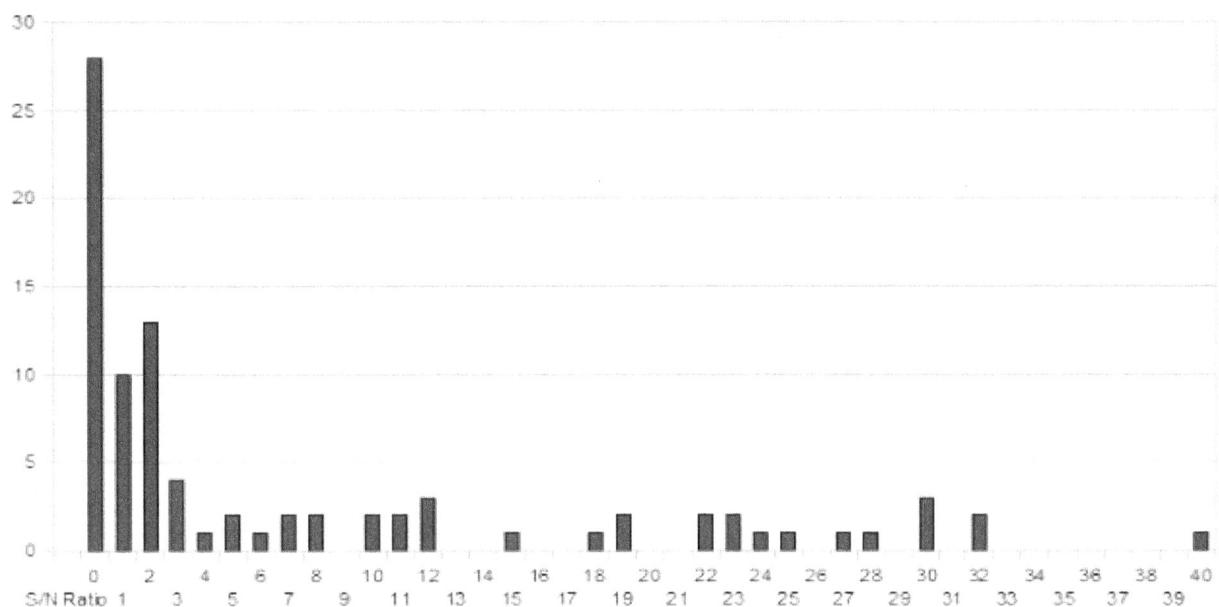

The PL-390, using Silicon Lab's Si4734 low-IF DSP chip, is superb at FM DXing. Above is a plot of number of stations heard (y-axis) versus signal-to-noise ratio in dB (x-axis). The PL-390's ETM (Easy Tune Memory) was able to find 33 stations. Hand searching revealed 16 more (total is 49). Another pass revealed 10 more (total is 59). Working dusk and dawn revealed 16 more for a total of 75 stations! A folded 5/8 wave dipole was built using 42 inches of 300-ohm twin-lead wire supported with a cardboard tube. The dipole is directional: rotating it in three dimensions allowed for 13 more stations to be identified. The PL-390 was able to pick up a total of 88 FM radio stations.

The PL-390 was extensively tested against the DE1102, SW7600GR, DE1103, and RP2100 portables. The DE1102 is not really in the same league as the other four radios; although, it does have decent FM. The SW7600GR is very sensitive: the DE1103 and RP2100 are as well. Where the PL-390 shined was at selectivity and adjacent channel rejection. Time and time again, when parked next to stronger stations, the other radios experienced blocking. Faint stations were wiped out by the energy of the adjacent station. And this was not an isolated occurrence like it is on AM. Viewing the chart above: 25 stations have a S/N ratio of at least 10 and 14 stations have a S/N ratio of at least 20. Weak stations sound better on the PL-390. This is not trivial as the other radios are quite capable on FM. The Achilles heal of the PL-390 was that only 23 stations were resolvable in stereo. Tecsun engineers may wish to alter the value of the property "FM_BLEND_STEREO_THRESHOLD".

The PL-390 was also much easier to use due to its ergonomics. The RP2100 has no keypad and is heavy. The DE1103 and DE1102 have no volume wheel. The DE1102 and SW7600GR have no tuning knob. If you love FM DX, Tecsun and Silicons Labs have produced a winner: the PL-390. FM (88-108 MHz) reception using Silicon Labs' low-IF DSP chip is clearly superior to its analog competitors.

3.16 Tecsun PL-390 Cheat Sheet
Using the PL-390.
©2011

GRAY indicates a button must be depressed and held.
OFF indicates that the radio must be off.
BOLD indicates use of a knob or jack.

Function	Sequence	Comments
POWER	POWER	Toggles On/Off.
KEY LOCK	DISPLAY (key)	Toggles On/Off.
BEEP	0 (BEEP On/Off)	OFF Toggles On/Off.
SMART LIGHT	FM ST (light bulb)	OFF Turns light off after 3 seconds.
LIGHT ON	LIGHT	Light on continuously.
TIME FORMAT	2 (12/24)	OFF AM/PM is 12; military is 24.
SET CLOCK	TIME **KNOB** TIME **KNOB** TIME	Or direct entry: 6:45 is entered as 0 6 4 5.
SLEEP TIMER	POWER LIGHT **KNOB**	OFF Hold LIGHT and turn knob.
BUZZER ALARM	BELL **KNOB** hour, minute.	BELL toggles On/Off.
RADIO ALARM	Use BELL as enter key.	RADIO for radio; M and RADIO sets station.
STOP ALARM	POWER	Press once for BELL; twice for RADIO.
SNOOZE	LIGHT	Snooze is 5 minutes.
USB CHARGE	M (battery)	OFF Toggles charging On/Off. Auto stops.
LINE IN	**LINE-IN-JACK**	>> on display. Plays audio through radio.
BCB STEPS	3 (9/10 kHz)	OFF In US 10/°F; elsewhere 9/°C.
LW BAND	MW/LW	OFF Allows access to LW: 153 to 513 kHz.
FM BAND	1 (FM SET)	OFF US/EU 87.5; Russia 64; Japan 76.
FM STEREO	FM ST	>||< on display.
DISPLAY	DISPLAY	Rotates: time, alarm, temperature, signal.
BAND	SW^{UP} or SW^{DOWN} or MW/LW or FM	SW by bands. MW/LW is a toggle.
FILTER	AM BW	Selects between 6, 4, 3, 2, and 1 kHz.
ETM	ETM	Easy Tuning Mode (finds stations).
START ETM	ETM	SW: 3.25 min. AM: 20 sec. FM: 7 sec.
STORE ETM	<frequency> ETM	Add a station to ETM.
MANUAL TUNE	VF **KNOB**	Turn slow for 1 kHz >; fast for 5 kHz >>.
AUTO BROWSE	VF VF	Browse 5 seconds. Press M to store.
DIRECT ENTRY	VF <frequency>	On FM: 98.6 is entered as 9 8 6.
ATS AUTOMATIC TUNE SCAN	SW^{DOWN} (all bands) or SW^{UP} (one band) or MW/LW or FM	Finds stations and replaces memories. Afterward, browse using the tuning knob. SW^{UP} only writes to blank memories.
STORE MEM	M M or M	Store memory.
MEMORY MODE	VM **KNOB**	Knob tunes through memories. DELETE DELETE deletes memories.
BROWSE MEM	VM VM	Browse 5 seconds. Press DELETE to delete memories.
RECALL MEM	VM <number> ENTER	Recall memory.
DELETE MEM	VM DELETE DELETE	Deletes one band. **ALL** on display.

3.17 GP-5/SSB
Silicon Labs' Game Changer
©2015

    The $85 CountyComm GP-5/SSB is a radio that covers 76 to 108 MHz FM and 150 to 29,999 kHz AM/SSB. With some wire clipped to its 18" whip, it performs like a PL-390. In 2011, I asked Silicon Labs to add algorithms to demodulate SSB. No SSB is a drawback of both the PL-390 and PL-380. The GP-5/SSB adds SSB and a BFO that tunes in 50-Hz, 20-Hz, and 10-Hz steps. This is as fine as the legendary Drake R8B. Like the ICF-SW7600GR, the GP-5/SSB can switch between USB and LSB. The radio comes with: a MW ferrite, manual, pouch, 150" wire antenna, and ear-buds.

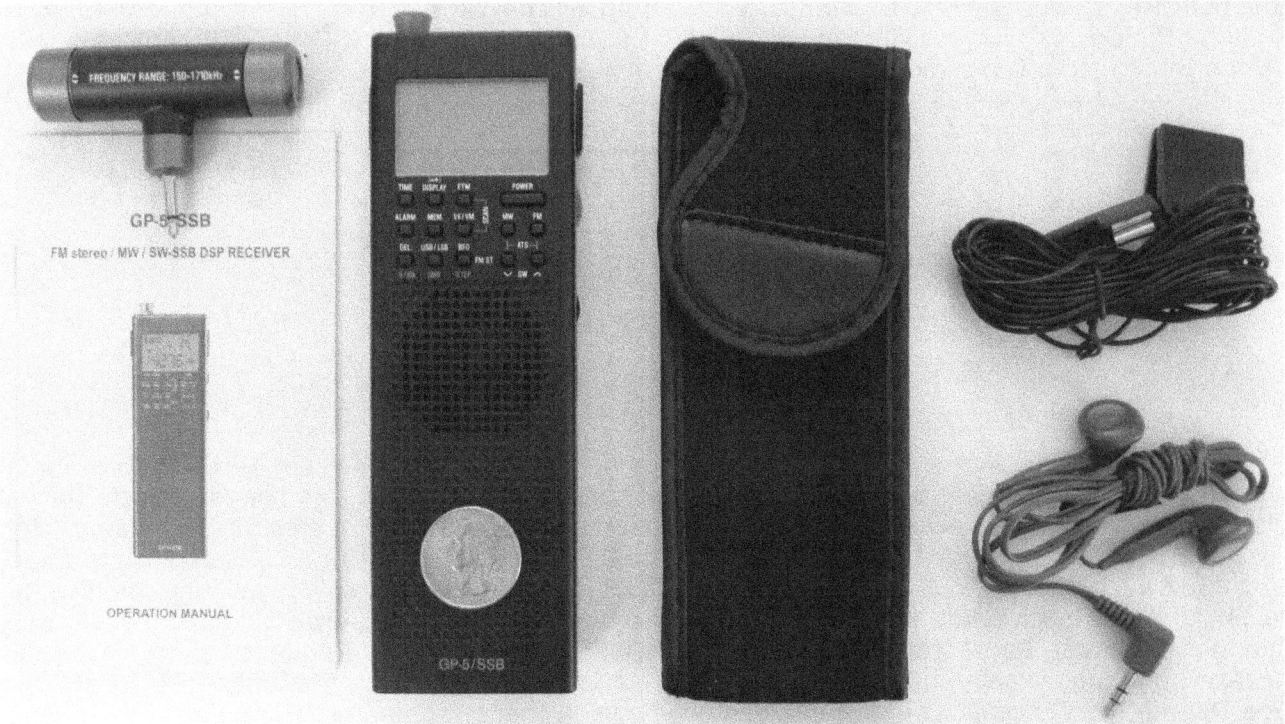

    First impressions: solidly build, sharp display, firm buttons, one handed use, antenna hard to extend. The GP-5/SSB displaces only 9.2 cubic inches. The "easy tuning mode" scans MW in 19 seconds, SW in 218 seconds, and FM in 13 seconds. CountyComm states that 3 "AA" batteries will last 225 hours at 40% volume. Current at zero volume is 23.4 mA (89.7 hours) and 31.9 mA with the backlight on (65.8 hours). The backlight automatically shuts off after 5 seconds. To remove the battery lid, push down on the grip dots, slide the lid rearward 5 millimeters, and then lift upward.

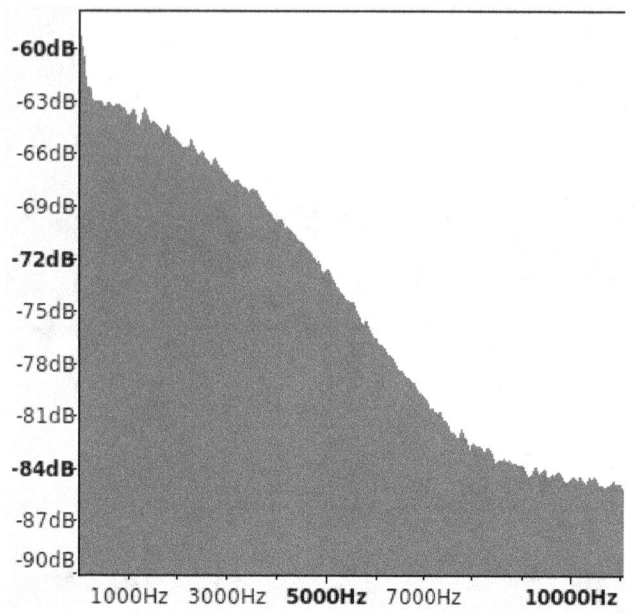 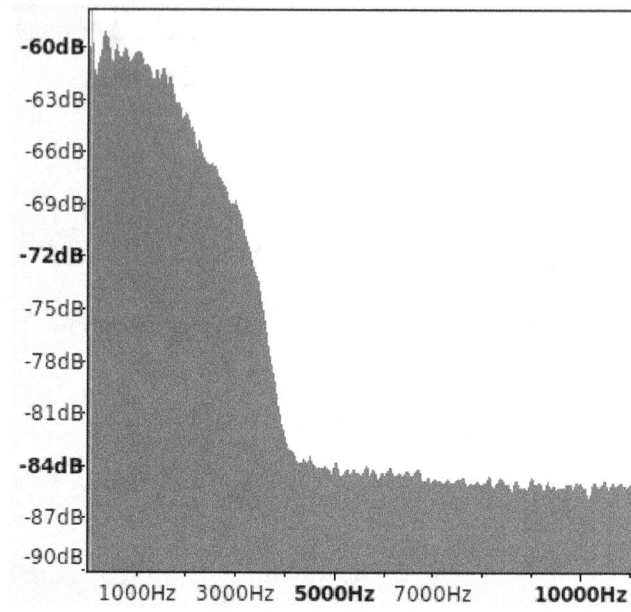

The bandwidth on AM (above, left) could be described as a gentle ~8-kHz (*4-kHz of audio*) filter: 2.25 shape factor. Good for AM listening. The bandwidth on SSB (above, right) could be described as a sharp ~2.5-kHz (*2.5-kHz of audio*) filter: 1.60 shape factor. Good for SSB work.

It takes 2 seconds to enter the SSB mode. The GP-5/SSB is *first-rate* at ECSS, or tuning AM stations in SSB mode. Without proper suppression of the opposite sideband, warble occurs. ECSS can be assisted via: analog filter (IC-R75, DE1103, KA1102), analog phasing (R8B, E1, ICF-SW7600GR), or digital filter (GP-5/SSB). ECSS using Silicon Labs' latest chip is superior to both analog solutions. In fact, the GP-5/SSB is better at ECSS than the $630 ICOM IC-R75. Analog filters are no match for a DSP's digital filters. The big thing missing on the GP-5/SSB is multiple filters. This may be an advantage in terms of simplicity because the AM and SSB filters are well chosen.

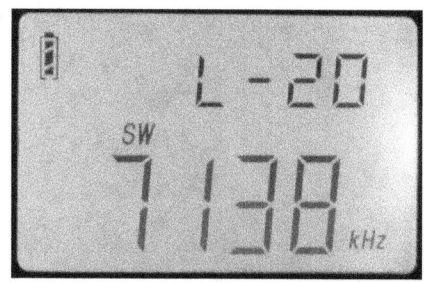

The GP-5/SSB was tested against the reigning portable SSB king: the DE1103. The audio was cleaner on the GP-5/SSB. The DE1103 takes finesse to get the audio to sound correct. This is also true of the ICF-SW7600GR and any other radio that uses a potentiometer for BFO tuning. The GP-5/SSB was much easier to tune on SSB and the frequency did not drift. Because there is no LSB/USB switch on the DE1103 (and other radios), users are left to properly offset tune the filter to minimize adjacent interference. The GP-5/SSB is proving Silicon Labs' newest chip to be a game changer. Look for future radios with multiple AM/SSB filters, direct frequency entry, a larger speaker, and longer antenna. The Tecsun PL-880, unfortunately, fails on SSB and lacks sensitivity. See Dave N9EWO's great review. Radio history is being made, again, by Silicon Labs' engineers. The age of analog radio is being phased out by a great American company in Austin, Texas.

3.18 RadioShack 12-586
Another winner; care of Silicon Labs.
©2013

1. INTRODUCTION

The $15 RadioShack AM/FM Pocket Radio, model 12-586, contains **Silicon Labs'** Si4820A mechanically-tuned, low-IF, DSP, radio chip. The radio takes up 13.6 cubic inches and weighs 170 grams, with 2 AA batteries. On the left is an on/volume pot. On the right a non-vernier tuning pot, AM/FM switch, and earphone jack. On top is a 16.25 inch FM antenna. Internally is a 2 inch ferrite (~71 turns litz). On back is a large battery door. A lanyard loop is included. The speaker is 8-ohm and 0.5 Watt. Current drain is 27.1 mA on AM and 23.2 mA on FM (speaker or earphone).

2. PRINTED CIRCUIT BOARD

The 12-586's one-sided, SMT PCB contains: 2 IC's, 2 diodes, 1 inductor, 17 capacitors, 14 resistors, and 2 zero-ohm jumpers. The component side includes: 5 electrolytic caps, 1 fork, 1 inductor, 2 pots, 1 ferrite, 1 band switch, and 1 earphone jack. The PCB date is 04-18-2012. IC1 reads: 4820A10GU (date code: week 36 of 2012). This is Silicon Labs' first generation chip; mono, no bass/treble, and no FM/tune LEDs. IC2 reads: TDA1822M, a dual power audio amplifier.

The PCB has solder balling. Two jumpers, to bridge ground planes, were omitted. Missing components included a cap for stereo output (*pin 23*) and a jumper/resistor setting volume mode. The original design may have used the second generation chip: Si4835. Hand soldering is sloppy.

The 12-586 is disassembled by removing: 2 exterior screws, a flat edge (screwdriver blade) to open the shell (battery end first), 2 PCB screws, and 1 screw on each of two pot wheels.

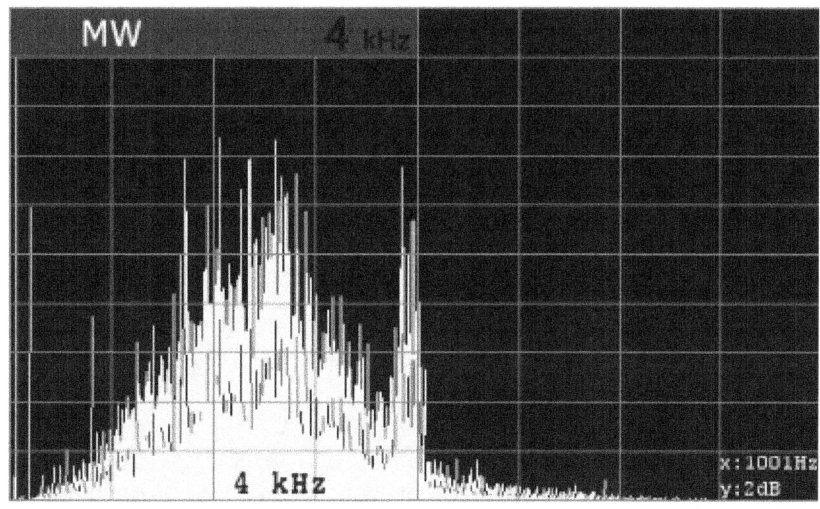

The graph above shows the attenuation of background noise via the 12-586's MW filter. Sample rate is 48,000 Hz. The x-axis is linear: 1001 Hz per division. The y-axis is log 10: 2 dB per division. The trace represents peeks over a timed period. This is a sharp 4-kHz audio filter.

3. IMPRESSION

I am a fan the Sony ICF-S10MK2; it is easy to hack. However, the 12-586 offers better out-of-the-box AM and FM reception. It also covers 1610 to 1710 kHz. The 12-586's Si4820 chip offers little to hack; outside of adding a fine tuning pot and voltage reading port to determine frequency. The 12-586's MW antenna is small but of good quality; this is a steal of a radio at $15.

4. UNOFFICIAL 12-586 BLOCK DIAGRAM

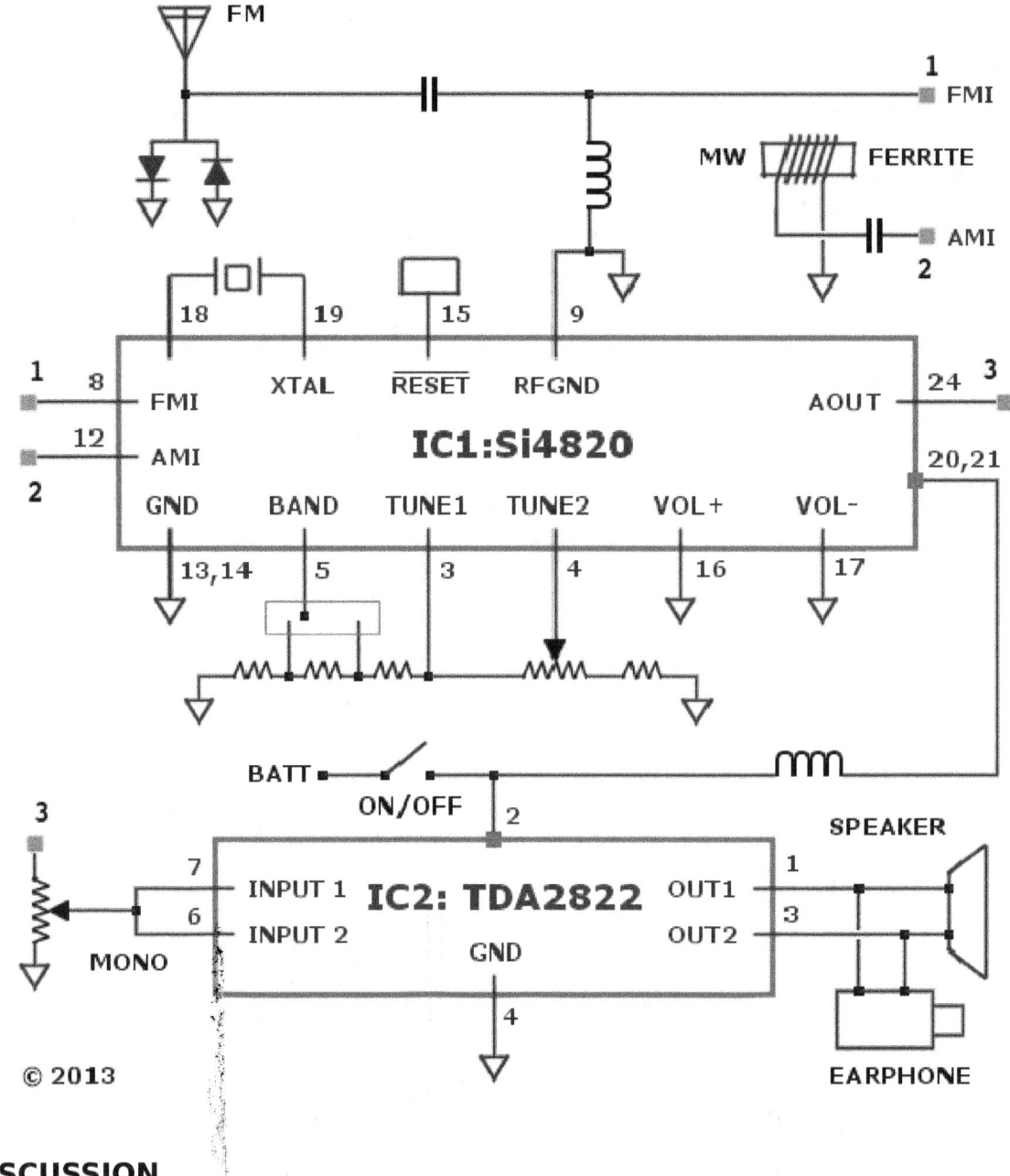

© 2013

5. DISCUSSION

Radio technology has steadily evolved. Tubes turned to transistors; turned to chips. LC oscillators turn to Phase Locked Loops; turned to Direct Digital Synthesis. Hole-through turned to surface-mount. LC filters turned to ceramic filters; and now DSP filters. Regens turned to single-conversion; turned double-conversion; and now low-IF. Silicon Labs is leading these changes. And we can experience a taste of that change by walking into our local RadioShack and shelling out a mere $15. We definitely live in remarkable times to have radio as a hobby.